KU-194-801

Two Grade Arithmetic

BOOK FOUR

By K. LOVELL, B.Sc., M.A., Ph.D.(Lond.)

Professor of Educational Psychology, University of Leeds

and C. H. J. SMITH, B.Sc., M.A., Ph.D.(Lond.)

formerly Senior Lecturer in Education and Methodology of Mathematics, Borough Road College

GINN AND COMPANY LTD

LONDON AND AYLESBURY

© K. Lovell and C. H. J. Smith 1956, 1969, 1971, 1975
3757603

Revised and reset 1969
First metric edition 1971
Revised metric edition 1975
Second impression 1976

Published by Ginn and Company Ltd.
Elsinore House, Buckingham Street, Aylesbury, Bucks HP20 2NQ

Product No. 112110738   ISBN 0 602 21885 3   (Pupils)
Product No. 112110827   ISBN 0 602 21889 6   (Answers)

Printed in Great Britain at the University Press, Oxford
by Vivian Ridler, Printer to the University

PREFACE

I hope that these books will give you experience of working some important kinds of examples. Those on the right hand page are usually a little harder than those on the left.

Your teacher will tell you which examples you should work. If you try hard and work steadily and neatly, I hope you will get right all the exercises that you attempt, and that you come to enjoy mathematics.

Leeds
1971

K. LOVELL

ADDITION AND SUBTRACTION

First try these

Set 1

1	2134 823 426 +4017	2	245 5132 49 +1326	3	1417 332 7149 +26	4	5247 191 36 +628
5	436 2157 49 +8903	6	6439 2073 9104 +4826	7	5 640 237 19 319 +5 400	8	14 759 20 128 97 +1 085
9	7 019 14 936 5 018 26 915 +1 342	10	3 714 67 391 5 160 32 718 +6 123	11	84 313 22 956 7 407 16 539 +3 803	12	35 142 62 573 3 804 1 937 +58 921

Set 2

1	6875 – 3429	2	13 416 – 9 843	3	24 017 – 15 928	4	35 981 – 9 467
5	43 106 – 29 195	6	50 372 – 31 468	7	59 632 – 10 859	8	72 403 – 59 728
9	80 013 – 55 275	10	76 182 – 9 784	11	93 652 – 37 419	12	100 025 – 18 649
13	121 075 – 82 157	14	151 286 – 45 974	15	130 175 – 68 038	16	107 510 – 59 284

ADDITION AND SUBTRACTION

Now try these

Set 1

1		2		3		4	
	3897		6884		9325		10 326
	54		209		4083		7 195
	1792		7536		1705		8 230
	+5516		+1007		+827		+9

5		6		7		8	
	12 513		33 196		29 354		825
	6 094		9 328		16 198		37 316
	15 012		4 155		8 009		25 944
	+85		+10 082		+7 328		+48 536

9		10		11		12	
	86 503		63 282		297 208		132 456
	27 431		175 306		35 819		452 803
	186		47 087		6 013		217 612
	20 243		11 203		364 254		63 097
	+9 357		+514		+7 909		+4 567

Set 2

1		2		3		4	
	7007		27 015		25 143		48 325
	−4986		−10 783		−19 899		−28 537

5		6		7		8	
	57 209		170 020		389 176		99 273
	−36 891		−59 360		−220 484		−73 567

9		10		11		12	
	621 309		739 451		175 604		204 836
	−353 875		−68 930		−95 736		−137 946

13		14		15		16	
	563 483		815 237		406 291		913 396
	−194 607		−208 108		−374 685		−496 497

MULTIPLICATION AND DIVISION

First try these

Set 1

1 38×17	**2** 43×29	**3** 52×27
4 36×35	**5** 57×72	**6** 68×45
7 89×26	**8** 74×41	**9** 93×99
10 88×93	**11** 56×84	**12** 65×56
13 185×19	**14** 287×23	**15** 346×35
16 461×29	**17** 318×74	**18** 429×63
19 584×71	**20** 647×98	**21** 705×86

Set 2

1 $1449 \div 21$	**2** $1936 \div 22$	**3** $2162 \div 23$
4 $2201 \div 31$	**5** $1792 \div 32$	**6** $3230 \div 34$
7 $2501 \div 41$	**8** $3910 \div 42$	**9** $3612 \div 43$
10 $4135 \div 51$	**11** $3423 \div 52$	**12** $5194 \div 53$
13 $3726 \div 61$	**14** $5830 \div 62$	**15** $3075 \div 25$
16 $4236 \div 37$	**17** $5946 \div 49$	**18** $4720 \div 46$
19 $7921 \div 68$	**20** $8326 \div 77$	**21** $6270 \div 57$
22 $9794 \div 83$	**23** $8647 \div 79$	**24** $8090 \div 65$
25 $1313 \div 101$	**26** $1905 \div 127$	**27** $3915 \div 135$

MULTIPLICATION AND DIVISION

Now try these

Set 1

1 65×29 **2** 78×41 **3** 83×52

4 98×69 **5** 243×35 **6** 487×58

7 697×46 **8** 718×67 **9** 860×73

10 985×95 **11** 897×84 **12** 1234×15

13 1362×24 **14** 1480×37 **15** 1571×43

16 1891×56 **17** 2050×69 **18** 2369×71

19 3985×84 **20** 4768×98 **21** 5007×79

Set 2

1 $1891 \div 31$ **2** $1664 \div 32$ **3** $3304 \div 32$

4 $3114 \div 41$ **5** $3572 \div 42$ **6** $4214 \div 49$

7 $2889 \div 57$ **8** $4095 \div 63$ **9** $6984 \div 72$

10 $8536 \div 78$ **11** $6601 \div 89$ **12** $8424 \div 54$

13 $9737 \div 91$ **14** $7812 \div 93$ **15** $9165 \div 65$

16 $12\,153 \div 52$ **17** $19\,856 \div 76$ **18** $27\,006 \div 85$

19 $31\,903 \div 71$ **20** $47\,948 \div 43$ **21** $53\,200 \div 26$

22 $73\,814 \div 34$ **23** $95\,219 \div 23$ **24** $3542 \div 154$

25 $10\,395 \div 297$ **26** $32\,400 \div 675$ **27** $61\,890 \div 897$

MONEY ADDITION AND SUBTRACTION

First try these

Set 1

1 £
  7·81½
  4·17
+2·06½

2 £
  23·62
  39·40
+2·98

3 £
  47·75
  35·61½
+18·20½

4 £
  65·13
  26·48
+10·79

5 £
  89·74
  2·33
+15·62

6 £
  101·27
  30·94
+45·85

7 £
  123·15
  91·26
+87·60½

8 £
  236·00
  59·18
+168·97

9 £
  371·28
  185·36
+107·49

Set 2

1 £
  8·35
−2·89

2 £
  20·68
−14·75½

3 £
  39·12
−28·96

4 £
  63·17
−49·82½

5 £
  98·22
−61·53

6 £
  137·41½
−58·60½

7 £
  170·93
−140·14

8 £
  253·16
−96·91

9 £
  318·54
−185·64

10 £
  395·68½
−288·57

11 £
  423·06
−395·18

12 £
  500·00
−417·63½

MONEY ADDITION AND SUBTRACTION

Now try these

Set 1

1 £
74·18½
68·29½
+95·83

2 £
123·64
6·80
+85·17

3 £
98·19½
100·37
+84·55½

4 £
139·47
125·01
+59·68

5 £
108·00½
326·21½
+153·75½

6 £
267·45½
140·39
+306·84

7 £
675·87
129·14½
+287·60

8 £
479·29½
350·51½
+806·48

9 £
683·36
509·58
+957·49

Set 2

1 £
50·36
−42·84½

2 £
65·27
−38·19½

3 £
103·01
−72·81

4 £
191·78
−125·69

5 £
216·53½
−98·57

6 £
476·51½
−352·39½

7 £
632·74
−497·85

8 £
950·10
−29·80½

9 £
1297·84
−846·75

10 £
1500·06
−1428·34½

11 £
1923·27½
−943·68

12 £
2860·54
−1895·76½

MONEY MULTIPLICATION AND DIVISION

First try these

Set 1

1 £
3·17
×5

2 £
8·29½
×7

3 £
12·38
×6

4 £
15·81½
×4

5 £
21·50
×8

6 £
19·13
×9

7 £
27·47
×3

8 £
10·65½
×11

9 £
13·28
×10

10 £
35·09½
×6

11 £
43·17
×5

12 £
56·83
×4

Set 2

1 £
3)19·47

2 £
7)38·01

3 £
5)92·68

4 £
4)125·16

5 £
6)100·34

6 £
8)140·72½

7 £
12)158·07

8 £
11)187·49½

9 £
10)206·13

10 £
9)221·61

11 £
7)246·38

12 £
12)312·09

MONEY MULTIPLICATION AND DIVISION

Now try these

Set 1

1. £
 4·98½
 ×9
———

2. £
 11·14½
 ×7
———

3. £
 20·62
 ×10
———

4. £
 42·83½
 ×8
———

5. £
 63·05
 ×12
———

6. £
 71·47
 ×11
———

7. £
 14·36
 ×13
———

8. £
 19·00½
 ×15
———

9. £
 30·51
 ×12
———

10. £
 13·94½
 ×20
———

11. £
 23·76
 ×11
———

12. £
 15·11
 ×9
———

Set 2

1. £
 8)49·16
———

2. £
 11)147·02½
———

3. £
 9)300·51
———

4. £
 12)473·92
———

5. £
 10)686·18
———

6. £
 7)952·31½
———

7. £
 14)12·03
———

8. £
 20)39·50
———

9. £
 13)98·67
———

10. £
 12)189·22½
———

11. £
 9)394·00
———

12. £
 11)873·60
———

BILLS

First try these

Find the total cost of

1 2¼ kg at 10p per kg
14 g at 3½p per half g
1½ tonnes at 75p per
100 kg

2 20 articles at 5 for 8p
120 articles at 1½p each
90 articles at 35p per
score

3 10 kg at 5½p per kg
5 g at 16p per g
½ kg at £1·80 per kg
2¼ kg at 2½p per ¼ kg

4 5½ litres at 3p per litre
6 litres at 24p per litre
17 litres at 2p per ½ litre
4 litres at 17p per ½ litre

5 4 m at 6½p per m
2 m at 7p per cm
15 at 36p per doz.
300 at 1p each

6 6 h at 16p per h
4 h at 21p per h
8 h at 25½p per h
5 h at 18p per h

7 10 pr gloves at 95p a pr
7 pr shoes at £6·50 a pr
9 pr sandals at £2·60 a pr
2 pr socks at 48p a pr

8 14 articles at 32p each
13 articles at 15p each
11 articles at 26p each
19 articles at 40p each

9 1½ tonnes at £5·00 per
tonne
10½ kg at 10p per kg
1¾ m at 36p per m
50 cm at 54p per m

10 3 rugs at £6·60 each
6 m carpet at £3·90 per
m
2 carpets at £45·00 each
4½ m carpet at £5·00 per
m

11 6 dresses at £11·00 each
4 pr shoes at £5·50 a pr
2 coats at £21·00 each
15 m cloth at £2·80 per m

12 3 doz. at £0·45 per doz.
16 doz. at 1p each
30 at 6 for 25p
5 doz. at £1·00 per score

BILLS

Now try these

Find the total cost of

1 16 m at 5½p per m
¾ kg at 1p per g
¼ kg at 2p per g
30 m at 10½p per ½ m

2 5 score plants at 48p per doz.
240 m at 17p per m
150 trees at £1·78 each
40 h labour at 84p per h

3 7 rolls wallpaper at 84p per roll
1½ litres paint at 96p per ½ litre
32 h labour at 80p per h
21 h at 25p per h

2750 bricks at £40·00 per 1000
¾ tonne at £15·60 per tonne
220 m at 18p per m
900 sq m at 60p per sq m

5 24 litres at 7p per litre
100 eggs at 48p per doz.
28 kg at 19p per kg
½ kg at 75p per kg

6 2½ doz. buns at 3 for 12p
6½ doz. cakes at 6p each
6 score at 35p per doz.
50 doz. at 10p per score

7 5 chairs at £19·52 each
4 tables at £25·28 each
6¼ m carpet at £6·00 per m
180 sq m at £1·25 per sq m

8 70 kg at 7 kg for 5p
2½ tonnes at £8·30 per tonne
3½ kg at 48p per kg
26 kg at 2 kg for 63p

9 25 at £9·00 per doz.
239 at 4p each
481 at 6p per doz.
13 at £7·20 per 12 doz.

10 250 at £20·00 for 10 000
1000 at 1p each
241 at 3½p each
78 at 60p a doz.

11 50 at £1·70 per 100
10 000 at £2·00 per 100
95 at 30p a doz.
1250 at £30·00 for 500

12 1 sq m at 1p per sq cm
44 h at 21p per h
½ kg at ½p per g
15 litres at 5½p per ½ litre

PROBLEMS

First try these

1 How much profit did a builder make if he built a house for £5984·79 and sold it for £7250·00?

2 Seven thousand two hundred blankets were packed, in equal numbers, in 96 bales. How many blankets were there in each bale?

3 Twelve excursion trains, each consisting of 9 carriages, left London for the coast. If each carriage held 48 people, how many passengers were carried?

4 If £63·56 is divided equally among 4 men and 3 women, find the share of each.

5 Find the cost of 4 tonnes at £15 per tonne, $7\frac{1}{2}$ tonnes at £17·00 per tonne, and $3\frac{1}{2}$ tonnes at £17·50 per tonne.

6 In three years a steamship travelled 52 075 sea miles, 47 896 sea miles and 51 294 sea miles. How far did it travel altogether in three years?

7 The sum of £2500 was needed to add a new wing to a school library. How much money was still required after the following collections had been made: £834·26, £100·50 and £790·35?

8 Find the number nearest to ten thousand which is exactly divisible by 29.

9 Nine similar articles were sold for £13·41. What was the price of each?

10 A hospital uses on an average 430 eggs per day. How many will it use during the month of May?

11 Three hundred workmen strike because an increase of 2p per h in pay is not granted. What extra sum of money would have been required, for a week of 40h, if this increase had been given?

12 A man bought 20 000 oranges and sold them all at a rate of 10 for 40p. How much money did he get for them?

13 A man bought 29 gross of eggs and sold 347 dozen. How many eggs had he left?

PROBLEMS

Now try these

1 Find the cost in pounds of 80 400 articles each costing one penny.

2 What is the difference between $(216 + 137) \times 53$ and $216 + (137 \times 53)$?

3 A merchant bought 56 boxes of eggs, with 120 in each. On finding that 5 eggs in each box were cracked, he repacked the good eggs in 40 boxes, putting the same number in each. How many eggs were packed in each box?

4 Find the difference between £87·36 + £14·90 + £270·81½ + £5·02½ and one half of a thousand pounds.

5 In a city there were 320 958 men, 331 716 women and 123 649 children under 16 years of age. How many people lived there?

6 A woman left £983·64 to be divided equally among 7 people. How much money did each receive?

7 How much money would a bus conductor take by collecting 105 2½p fares, 71 3p fares, 64 3½p fares and 29 5p fares?

8 Find the total cost of
   240 at 1½p each
   480 at 2½p each
   36 at 33p each
   960 at 2p each

9 Find the total cost of
   24 litres at 5p per ½ litre
   5 kg at 7p per 100 g
   28 m at 4½p per ½ m
   42 h at 39½p per h

10 Divide the difference between 72 427 and 90 005 by 34.

11 Tom's elder brother goes to work. Each week his bus fares cost him £0·90, his midday meals cost £2·00 and he pays £8·50 at home. How much does he spend on these items in four weeks?

12 Tom's father earns £168·00 per month. He spends ¼ of this on rent and ⅖ on food. How much more does he spend on food than on rent in 12 months?

13 From the largest number you can make with the numbers 6, 8, 3, 5, 9, take the smallest you can make with the same numbers and multiply your result by 2.

USING DIRECTED NUMBERS

First try these

$$-7 \quad -6 \quad -5 \quad -4 \quad -3 \quad -2 \quad -1 \quad 0 \quad +1 \quad +2 \quad +3 \quad +4 \quad +5 \quad +6 \quad +7$$

Set 1

1 $^+2 + {}^+3 =$

2 $^+4 + {}^+7 =$

3 $^+1 + {}^+6 + {}^+4 =$

4 $^+8 + {}^+2 + {}^+3 + {}^+5 =$

5 $^+20 + {}^+5 + {}^+16 =$

6 $^+13 + {}^+21 + {}^+9 =$

7 $^+10 + {}^+15 + {}^+17 + {}^+6 =$

8 $^+9 + {}^+14 + {}^+18 + {}^+20 =$

Set 2

1 $^+10 - {}^+9 =$

2 $^+8 - {}^+4 =$

3 $^+13 - {}^+7 =$

4 $^+17 - {}^+12 =$

5 $^+21 - {}^+15 =$

6 $^+30 - {}^+23 =$

Set 3

1 $^+4 + {}^+3 - {}^+2 =$

2 $^+9 + {}^+11 - {}^+3 =$

3 $^+12 - {}^+10 + {}^+1 =$

4 $^+17 - {}^+8 + {}^+4 =$

5 $^+30 + {}^+9 - {}^+18 =$

6 $^+27 - {}^+19 - {}^+2 =$

7 $^+40 - {}^+13 - {}^+23 =$

8 $^+35 - {}^+11 - {}^+15 =$

USING DIRECTED NUMBERS

Now try these

Set 1

1 $^+12 + ^-5 =$
2 $^+10 + ^-7 =$
3 $^+18 + ^-14 =$
4 $^+23 + ^-15 =$
5 $^+16 + ^+9 + ^-20 =$
6 $^+30 + ^-28 + ^+6 =$

Set 2

1 $^-6 + ^+2 =$
2 $^-8 + ^+3 =$
3 $^+10 + ^-12 =$
4 $^+21 + ^-31 =$
5 $^+5 + ^+4 + ^-13 =$
6 $^+10 + ^-16 + ^+4 =$

Set 3

1 $^-5 + ^-3 =$
2 $^-7 + ^-8 =$
3 $^-12 + ^-6 =$
4 $^-14 + ^-10 =$
5 $^-20 + ^-15 =$
6 $^-17 + ^-26 =$

Set 4

(a) $^+6 - ^+8 =$
(b) $^-4 - ^-2 =$
(c) $^+9 - ^+12 =$
(d) $^-7 - ^-6 =$
(e) $^+18 - ^+30 =$
(f) $^-14 - ^-9 =$
(g) $^+25 - ^+42 =$
(h) $^-32 - ^-15 =$
(i) $^+50 - ^+61 =$
(j) $^-29 - ^-10 =$
(k) $^+38 - ^+17 =$
(l) $^-19 - ^-24 =$

REDUCTION AND QUOTITION

First try these

Reduction: change the following as indicated

1 £0·37 to halfpence
2 £0·28½ to halfpence
3 £4·08 to pence
4 393 pence to pounds
5 5 km 19 m to m
6 10 m 8 cm to cm
7 294 m to cm
8 12 days 18 h to h
9 402 sec to min and sec
10 8 kg 6 g to g
11 5 tonnes to kg
12 15 kg to g
13 129 h to days and h
14 3600 m to km and m
15 2 weeks 5 days to h
16 8000 g to kg
17 3¼ kg to g
18 745 g to kg
19 4¾ km to m
20 9050 cm to m

Quotition: find how many times

1 21p is contained in 84p
2 £2·05 is contained in £10·25
3 80 g is contained in 4 kg
4 61 m is contained in 366 m
5 1000 cm is contained in 1 km
6 5 km 250 m is contained in 31½ km
7 150 g is contained in 3 kg
8 2 min 20 sec is contained in 18 min 40 sec
9 1 day 18 h is contained in 8 days 18 h
10 22 litres is contained in 550 litres
11 24 half-litres is contained in 108 litres
12 37½p is contained in £3·00

REDUCTION AND QUOTITION

Now try these

Reduction: change the following as indicated

1 £0·95½ to halfpence   2 £0·87½ to halfpence
3 £10·07 to pence   4 1748 pence to pounds
5 19 m to cm   6 17 880 m to km and m
7 20 km to m   8 2 year 261 days to days
9 10 weeks 4 days to days 10 507 mo to yr and mo
11 18 tonnes 17 kg to kg   12 9¾ kg to g
13 2256 kg to tonnes and kg 14 1975 g to kg and g
15 19 765 m to km   16 3076 cm to m
17 1·238 kg to g   18 0·743 km to m
19 450 litres to half-litres 20 1000 ml to litres

Quotition: Find how many times

1 48p is contained in £7·20
2 £0·62½ is contained in £10
3 98 cm is contained in 588 cm
4 2 days 19 h is contained in 50 days
5 3 weeks 4 days is contained in 20 weeks 5 days
6 3¼ tonnes is contained in 195 tonnes
7 1½ m is contained in 3 km
8 2500 g is contained in 20 kg
9 11½ litres is contained in 57½ litres
10 7½ kg is contained in 30 tonnes
11 10 m 10 cm is contained in 10 km 100 m
12 9·5 cm is contained in 4¾ m

REFLECTION, ROTATION, TRANSLATION

All can try these

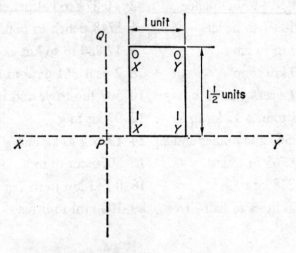

1 Draw the rectangle after:
  (*a*) Reflection in PQ
  (*b*) Reflection in XY
  (*c*) Rotation through 90° in a clockwise direction, about the centre of the rectangle
  (*d*) Rotation through 45°, in a clockwise direction, about the centre of the rectangle
  (*e*) Translation along XY in the direction of Y
  (*f*) Reflection in PQ followed by reflection in XY, followed by rotation through 90° in an anticlockwise direction about the centre of the rectangle

2 (*a*) Draw another rectangle doubling the length of the sides of the rectangle above
  (*b*) Find the ratio of (*i*) the perimeters and (*ii*) the areas of this larger rectangle and the rectangle above

MORE ABOUT SHAPES

All can try these

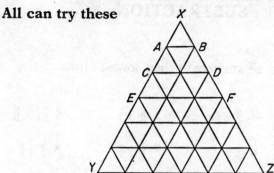

1 XYZ is an equilateral triangle of size 6 units.

  (*i*) How many small triangles are there above (*a*) AB, (*b*) CD, (*c*) EF?

  (*ii*) What have you noticed about the number of small triangles above a given line?

  (*iii*) How many small triangles are there above the line YZ? (Do not count)

  (*iv*) If the large triangle had a side of (*a*) 8 units, (*b*) 11 units, (*c*) 15 units, how many small triangles would there be above the base of the large triangle?

2

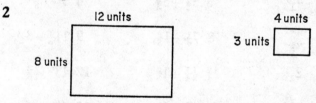

  (*i*) What is the ratio of the perimeter of the larger rectangle to that of the smaller?

  (*ii*) What is the ratio of the area of the larger rectangle to that of the smaller?

3 The length of the sides of a larger square are (*a*) 4 times and (*b*) N times those of a smaller square. In each case find the ratio of the areas of the larger and smaller squares.

FRACTIONS ADDITION AND SUBTRACTION

First try these

Write all answers in their lowest terms

Set 1

1 $\frac{1}{3} + \frac{1}{12}$ **2** $\frac{3}{4} + \frac{5}{16}$ **3** $\frac{2}{3} - \frac{1}{6}$ **4** $\frac{15}{16} - \frac{7}{8}$

5 $\frac{13}{18} - \frac{2}{9}$ **6** $\frac{3}{8} + \frac{5}{12}$ **7** $\frac{1}{6} + \frac{7}{9}$ **8** $\frac{2}{3} + \frac{4}{5}$

9 $\frac{7}{12} - \frac{1}{8}$ **10** $\frac{7}{9} - \frac{1}{2}$ **11** $\frac{7}{10} - \frac{2}{5}$ **12** $\frac{3}{4} - \frac{1}{6}$

13 $\frac{1}{2} + \frac{1}{3} + \frac{1}{4}$ **14** $\frac{1}{2} + \frac{3}{4} - \frac{5}{8}$ **15** $\frac{1}{3} + \frac{1}{6} - \frac{1}{9}$

16 $\frac{1}{4} + \frac{1}{6} + \frac{5}{8}$ **17** $\frac{5}{6} + \frac{2}{3} - \frac{1}{2}$ **18** $\frac{1}{4} - \frac{1}{8} + \frac{1}{6}$

Set 2

1 $7\frac{1}{2} + 4\frac{3}{4}$ **2** $5\frac{4}{9} + 1\frac{2}{3}$ **3** $1\frac{1}{3} + 2\frac{5}{6}$

4 $2\frac{4}{7} + 4\frac{3}{14}$ **5** $5\frac{1}{2} + 5\frac{2}{3}$ **6** $9 - 6\frac{5}{8}$

7 $2\frac{3}{4} - \frac{4}{5}$ **8** $7\frac{1}{6} - 1\frac{2}{3}$ **9** $11\frac{1}{8} - 3\frac{1}{24}$

10 $5\frac{3}{5} - 2\frac{9}{10}$ **11** $\frac{17}{24} + 1\frac{5}{8}$ **12** $13 - \frac{13}{16}$

13 $1\frac{4}{5} + 7\frac{11}{20}$ **14** $1\frac{3}{8} - \frac{4}{5}$ **15** $6\frac{3}{4} - 5\frac{7}{12}$

16 $1\frac{1}{8} - \frac{11}{12}$ **17** $8\frac{3}{4} - 1\frac{3}{20}$ **18** $2\frac{1}{4} + 6\frac{2}{7}$

19 $17\frac{3}{11} - 13\frac{7}{22}$ **20** $8\frac{5}{6} + 5\frac{1}{8}$ **21** $4\frac{3}{11} + 1\frac{1}{2}$

22 $5\frac{7}{16} - 2\frac{5}{8}$ **23** $7\frac{5}{12} - 2\frac{2}{5}$ **24** $9\frac{3}{8} + 1\frac{1}{9}$

FRACTIONS ADDITION AND SUBTRACTION

Now try these

Write all answers in their lowest terms

Set 1

1. $\frac{2}{3} - \frac{7}{18}$ 2. $\frac{9}{20} - \frac{7}{30}$ 3. $\frac{11}{16} - \frac{5}{24}$ 4. $\frac{4}{9} + \frac{1}{6}$

5. $\frac{7}{12} + \frac{5}{9}$ 6. $\frac{11}{14} + \frac{3}{7}$ 7. $\frac{4}{5} + \frac{3}{4}$ 8. $\frac{11}{12} - \frac{3}{8}$

9. $\frac{11}{15} + \frac{1}{3} - \frac{3}{5}$ 10. $\frac{3}{4} - \frac{3}{8} - \frac{1}{3}$ 11. $\frac{9}{10} - \frac{1}{5} - \frac{2}{3}$

12. $\frac{7}{8} - \frac{1}{3} + \frac{11}{12}$ 13. $\frac{15}{16} - \frac{3}{4} - \frac{1}{8}$ 14. $\frac{13}{20} + \frac{3}{5} - \frac{2}{3}$

15. $\frac{4}{5} + \frac{11}{30} + \frac{1}{2}$ 16. $\frac{11}{14} + \frac{1}{4} - \frac{3}{8}$ 17. $\frac{17}{18} - \frac{1}{9} - \frac{1}{12}$

Set 2

1. $\frac{8}{9} + 3\frac{1}{12}$ 2. $6\frac{4}{7} + 7\frac{5}{6}$ 3. $1\frac{2}{15} + 4\frac{11}{12}$

4. $13\frac{7}{10} + 2\frac{1}{15}$ 5. $9\frac{1}{4} - 5\frac{5}{6}$ 6. $11\frac{1}{6} - 4\frac{7}{9}$

7. $8\frac{5}{12} - 7\frac{13}{15}$ 8. $14\frac{9}{10} - 12\frac{2}{3}$ 9. $13\frac{2}{5} - \frac{9}{10}$

10. $3\frac{1}{3} + 4\frac{1}{6} + 2\frac{5}{9}$ 11. $3\frac{7}{9} - 1\frac{11}{18} - \frac{1}{3}$ 12. $6\frac{2}{7} + 8\frac{3}{4} - 2\frac{1}{2}$

13. $5\frac{13}{16} + 2\frac{1}{3} - \frac{7}{8}$ 14. $6\frac{5}{12} + 1\frac{1}{15} + 4\frac{1}{2}$ 15. $9\frac{1}{6} - 4\frac{1}{8} - 2\frac{1}{12}$

16. $8\frac{3}{4} - 2\frac{1}{7} + 1\frac{1}{2}$ 17. $10\frac{11}{15} + 1\frac{2}{3} - 6\frac{1}{6}$ 18. $\frac{8}{9} + 7\frac{1}{8} + 11\frac{1}{2}$

19. $10\frac{1}{4} - 6\frac{1}{5} + 1\frac{1}{6}$ 20. $12\frac{11}{14} - 2\frac{1}{4} - 1\frac{3}{8}$ 21. $6\frac{1}{7} - 2\frac{5}{6} - 1\frac{1}{3}$

FRACTIONS MULTIPLICATION AND DIVISION

First try these

Set 1

1 $\frac{2}{3}$ of 3

2 $2\frac{1}{6} \times 12$

3 $\frac{4}{9} \times 6\frac{3}{4}$

4 $3\frac{3}{5} \times 2\frac{1}{3}$

5 $\frac{7}{8} \times 5\frac{1}{3}$

6 $1\frac{5}{16} \times 2\frac{2}{7}$

7 $3\frac{3}{10}$ of $1\frac{4}{11}$

8 $5\frac{1}{4} \times \frac{8}{15}$

9 $4\frac{5}{6} \times 2\frac{4}{7}$

10 $7\frac{1}{7} \times 2\frac{1}{10}$

11 $\frac{5}{8} \times \frac{10}{11}$

12 $\frac{19}{35}$ of $\frac{25}{38}$

13 $\frac{10}{11} \div 5$

14 $\frac{7}{12} \div \frac{1}{4}$

15 $\frac{1}{3} \div 3\frac{1}{3}$

16 $9\frac{3}{5} \div 8$

17 $\frac{6}{11} \div \frac{3}{8}$

18 $4\frac{1}{3} \div 2\frac{1}{6}$

19 $2\frac{4}{5} \div 2\frac{2}{3}$

20 $3\frac{2}{11} \div 5\frac{1}{4}$

21 $1\frac{1}{7} \div \frac{24}{49}$

22 $6\frac{2}{9} \div 9\frac{1}{3}$

23 $\frac{17}{63} \div 1\frac{5}{12}$

24 $2\frac{7}{10} \div 13\frac{1}{2}$

Set 2

1 $\frac{3}{4} \times \frac{2}{9} \times \frac{3}{8}$

2 $\frac{2}{11}$ of $5\frac{1}{2} \times \frac{1}{6}$

3 $\frac{1}{3} \times \frac{3}{5} \times 2\frac{1}{7}$

4 $3\frac{1}{5} \times 1\frac{1}{8} \times 8\frac{1}{3}$

5 $16 \times 4\frac{3}{8} \times \frac{3}{5}$

6 $6\frac{3}{4} \times 2\frac{1}{9} \times \frac{2}{19}$

7 $(3\frac{1}{3} \times 1\frac{1}{2}) \div 7\frac{1}{2}$

8 $(\frac{11}{14}$ of $2\frac{4}{5}) \div \frac{3}{4}$

9 $(\frac{1}{11}$ of $5\frac{1}{2}) \div 6\frac{2}{3}$

10 $18 \div (\frac{9}{10}$ of $4\frac{2}{7})$

11 $6\frac{1}{4} \div (7\frac{1}{5}$ of $\frac{5}{18})$

12 $\frac{4}{13} \div (1\frac{3}{11}$ of $6\frac{2}{7})$

13 $(5\frac{5}{8} \div 3\frac{3}{4}) \times \frac{1}{12}$

14 $(\frac{13}{16} \div \frac{26}{49}) \times \frac{8}{21}$

15 $\frac{5}{24} \div (\frac{3}{8}$ of $\frac{7}{9})$

FRACTIONS MULTIPLICATION AND DIVISION

Now try these

Set 1

1 $1\frac{3}{13} \times 3\frac{9}{10}$

2 $3\frac{3}{8}$ of $1\frac{1}{9}$

3 $9\frac{3}{4} \times 3\frac{1}{13}$

4 $\frac{27}{49} \times \frac{28}{45}$

5 $8\frac{2}{5} \times 4\frac{1}{6}$

6 $6\frac{2}{7} \times 5\frac{8}{11}$

7 $\frac{9}{16} \times 1\frac{5}{7}$

8 $9\frac{5}{8}$ of $\frac{2}{21}$

9 $\frac{11}{18} \times 6\frac{3}{7}$

10 $\frac{5}{14}$ of $3\frac{1}{16}$

11 $4\frac{2}{9} \times \frac{3}{19}$

12 $6\frac{1}{4} \times 5\frac{1}{5}$

13 $4\frac{1}{4} \div 1\frac{1}{2}$

14 $13\frac{1}{2} \div 7\frac{5}{7}$

15 $30\frac{1}{4} \div 5\frac{1}{2}$

16 $2\frac{1}{4} \div 10\frac{1}{2}$

17 $2\frac{7}{9} \div 9\frac{1}{6}$

18 $\frac{21}{32} \div \frac{35}{48}$

19 $14\frac{2}{3} \div 9\frac{3}{7}$

20 $11\frac{2}{3} \div 13\frac{3}{4}$

21 $18\frac{1}{2} \div 8\frac{2}{9}$

22 $16\frac{1}{5} \div 11\frac{1}{4}$

23 $\frac{17}{18} \div 7\frac{5}{9}$

24 $5\frac{7}{19} \div 6\frac{3}{8}$

Set 2

1 $3\frac{3}{11} \times 4\frac{2}{5} \times \frac{1}{9}$

2 $\frac{11}{45} \times \frac{16}{33} \times \frac{5}{8}$

3 $8\frac{3}{4} \times \frac{2}{15} \times \frac{1}{7}$

4 $(4\frac{1}{5} \div 10\frac{1}{2}) \times \frac{3}{8}$

5 $\frac{9}{10} \times (6\frac{2}{9} \div 5\frac{3}{5})$

6 $\frac{7}{22} \times (1\frac{1}{17} \div 11\frac{5}{11})$

7 $3\frac{3}{4} \times 7\frac{11}{15} \times 2\frac{2}{3}$

8 $4\frac{2}{9} \div (6\frac{1}{4}$ of $2\frac{8}{15})$

9 $(11\frac{2}{5} \div 1\frac{7}{12}) \times 13\frac{1}{3}$

10 $14\frac{1}{6} \div (3\frac{1}{11}$ of $13\frac{3}{4})$

11 $5\frac{5}{8} \times 10\frac{1}{9} \times \frac{2}{13}$

12 $12\frac{1}{10} \times (\frac{7}{32} \div 15\frac{2}{5})$

13 $9\frac{1}{4} \div (17\frac{1}{2}$ of $10\frac{4}{7})$

14 $(8\frac{1}{3} \times 7\frac{7}{10}) \div 99$

15 $(\frac{13}{30} \div 3\frac{7}{15}) \times \frac{8}{9}$

PROBLEMS

First try these

1 Twenty six telephone posts, placed at equal distances, extend one km. Find the distance between two posts.

2 Express $2\frac{1}{2}$ litres as a fraction of $6\frac{1}{4}$ litres.

3 Our train took $2\frac{1}{4}$ h from station A to station B. There we waited $\frac{1}{5}$ h for another train. From station B to station C took $\frac{1}{2}$ h. How long were we on the journey from A to C? (Give the answer in h and min.)

4 How many hours are there between midnight on March 30th and midday on April 1st? How many days is this?

5 An article costs 15p. How many can be bought for $\pounds\frac{3}{4}$?

6 By how much is 5 kg 427 g greater than 3 kg 681 g?

7 A man paid $\frac{2}{3}$ of his bill in notes, $\frac{1}{4}$ in 50p pieces and the rest in 1p pieces. What fraction of the bill was paid in 1p pieces?

8 Find the total cost of 1000 articles at 2p each.

9 A bucket holds $24\frac{1}{2}$ litres and a watering-can 8 litres. Both are filled 6 times, and the water is poured into an empty tank. How much water is there then in the tank?

10 A large wheel revolves once in 1 min 15 sec. How many times will it revolve in $1\frac{1}{2}$ h?

11 A wholesale grocer bought 1 tonne of sugar. He sold 150 kg to one customer, 380 kg to another, and 275 kg to a third. How much sugar remained unsold?

12 Our garden fence is 36 m long. One-eighth of it was blown down in a gale. What length remained standing.

13 A man had $\pounds3400\cdot00$. He bought a car costing $\pounds655$, and spent $\frac{2}{9}$ of the remaining money. How much money had he then?

14 How many articles, each costing 39p, can be bought for $\pounds36\cdot27$?

15 How many times is 10 km contained in $92\frac{1}{2}$ km?

16 A coal merchant had 1500 kg of a special kind of fuel weighed up in 25 kg bags. How many bags were there?

PROBLEMS

Now try these

1 A bicycle was priced at £31·80, but was actually sold for £27·40. What was the reduction in price?

2 The area of a rectangle is $83\frac{3}{10}$ sq cm. Its width is $8\frac{1}{2}$ cm. What is its length?

3 If the sun rises at 04:05 and sets at 20:02, for how many hours and minutes is it above the horizon?

4 How many books costing £0·35 can be bought with £15·00 and how much money will be left over?

5 A milkman sells 132 half-litre, 244 litre and 64 quarter-litre bottles of milk in one day. If he sells it at 10p per litre, what are his takings?

6 Four litres of a liquid weigh 4 kg. Find the weight of a tank holding 560 litres.

7 Add $\frac{7}{16}$ to the difference between $9\frac{3}{16}$ and $5\frac{5}{8}$.

8 An avenue of trees is 900 m long and the trees are planted on both sides, $4\frac{1}{2}$ m apart. How many trees are there in the avenue, including those at each end?

9 How many 120 g packets of tea can be made from two chests, one holding 30 kg and the other 15 kg?

10 A factory burnt $3\frac{3}{5}$ tonnes of coal one month and $\frac{2}{3}$ of this amount the next month. How much coal did it use (a) during the second month, (b) during both months?

11 A train travelling at a steady speed covers 1 km in 42 sec. If it continues at this speed, how far will it go in 28 min?

12 How many m of cloth will be required for $1\frac{1}{4}$ doz. garments, if each needs $\frac{4}{5}$ m?

13 What must $6\frac{1}{3}$ be divided by to make $7\frac{1}{3}$?

14 The petrol tank of a motor car is $\frac{3}{5}$ full when it holds 30 litres. How many litres does it hold when full?

15 To the sum of 3 m 10 cm and 1 m 82 cm, add their difference. Divide the result by 2.

16 How many m of material costing £0·56 per m can be bought with £12·88?

DRAWING TO SCALE (1)

Plan of a Netball Court

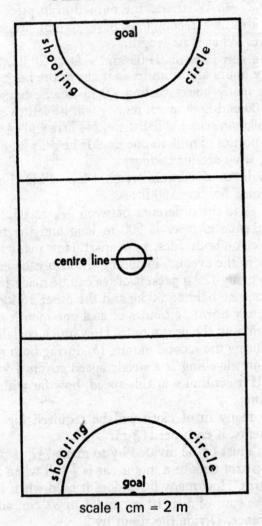

scale 1 cm = 2 m

Study this diagram, and answer the questions on the opposite page.

All can try these

1 What length of court does $\frac{1}{10}$ cm represent?

2 Draw lines, to this scale, to stand for 6, 10, 16, 5, 7, 11 m.

3 What length of line, drawn to this scale, stands for 4, 14, 26, 32, 40, 48, 56, 62, 74, 94, 100, 124 m?

4 What is the length and width of the court?

5 How far is the goal from the centre point of the field?

6 Find the radius of the shooting circle.

7 What is the length of the centre line?

8 How far is it from one corner of the court to the opposite corner, that is, across the diagonal of the court?

9 What is the area of the court?

10 How far from the centre line are the goal lines drawn?

11 Find the distance from the goal to one of the further corners.

12 What is the distance between the two points where the shooting circles come closest together?

13 Draw a rectangle 14 m by 12 m to the same scale, and find the distance between opposite corners in the original rectangle.

14 Draw, to the same scale, a square that has an area of 144 sq m. Then find the length of a diagonal. (Hint: first find the number which multiplied by itself gives 144.)

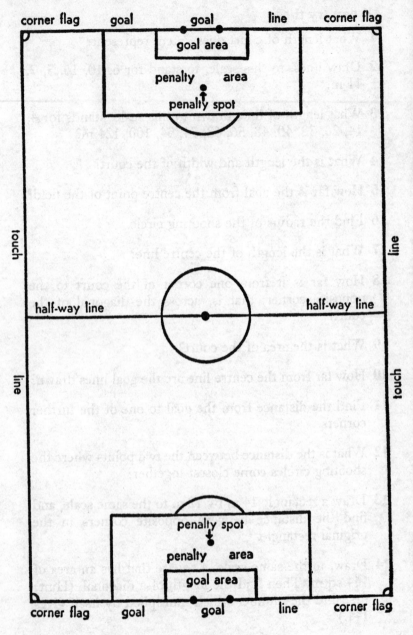

corner flag   goal   goal   line   corner flag

goal area

penalty   area

penalty spot

touch

line

half-way line   half-way line

line

touch

penalty spot

penalty   area

goal area

corner flag   goal   goal   line   corner flag

scale 1 cm = 6 m

DRAWING TO SCALE (2)

All can try these
Plan of a Football Pitch

1 What length on the field does (a) $\frac{1}{10}$ cm represent, (b) $\frac{5}{10}$ cm represent?

2 Draw lines to this scale to stand for 12, 18, 24, 42, 54, 66 m.

3 What length of line drawn to this scale stands for 9, 12, 27, 36, 78, 84, 90, 102, 114 yd?

4 What is the length and breadth of the field of play?

5 Find the length of the lines around (a) the goal area, and (b) the penalty area.

6 What is the distance from a corner flag to the nearest goal post?

7 Find the width of a goal.

8 Find the radius of the circle in the centre of the pitch.

9 Find the distance (a) from a penalty spot to the midpoint of the goal, and (b) between the two penalty spots.

10 What is the radius of the part of the circle which has the penalty spot as centre?

11 How far is the edge of the penalty area from the goal line?

12 Find the distance between the centre of the goal and (a) the furthest away corner of the goal area, and (b) the furthest away corner of the penalty area.

13 What is the distance between two opposite corner flags, that is, the length of a diagonal of the pitch? (Answer correct to the nearest m.)

14 Find the distance from a penalty spot to one of the nearer corner flags.

15 How far is it from a goal-post (**not** the centre of the goal) to the centre point of the nearer touch line?

16 What is the area of (a) the whole football pitch, (b) the goal area, (c) that part of the pitch outside the penalty areas?

AREA

First try these

Make a rough drawing in Nos. 3 and 4 and put in the measurements.

1 Find the area and the perimeter of each of the following squares or rectangles:
   - (a) Length 4 m   Breadth $2\frac{3}{4}$ m
   - (b) Length $2\frac{1}{2}$ m   Breadth $1\frac{1}{4}$ m
   - (c) Length $3\frac{1}{2}$ m   Breadth $3\frac{1}{2}$ m
   - (d) Length 15 cm   Breadth $6\frac{3}{4}$ cm
   - (e) Length 18 cm   Breadth $4\frac{1}{2}$ cm

2 Find the missing measurement (length or breadth) of the following rectangles, given the area and one side:
   - (a) Area 90 sq cm   Length 10 cm
   - (b) Area 156 sq cm   Breadth 12 cm
   - (c) Area 480 sq m   Length 60 m
   - (d) Area 900 sq m   Length $22\frac{1}{2}$ cm
   - (e) Area 5 sq m   Length 250 cm

3 A square carpet of side 3 m lies in the centre of a room 5 m × 4 m. What area of floor is uncovered, and what are the widths of the borders?

4 A small flag measures $\frac{3}{5}$ m × $\frac{1}{2}$ m. Find its area in (a) sq m, and (b) sq cm.

5 In the following figures the outside square has a side of 15 cm. Find the area of the shaded part of each figure.

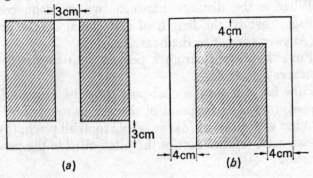

(a)   (b)

AREA

Now try these

Make a rough drawing whenever you can, and put in the measurements.

1 How many square metres of wood are required for a kitchen shelf if it is to be $2\frac{1}{2}$ m long and 50 cm wide?

2 A square field has a side of 75 m. What is its area? If you start from one corner and walk round the edge of the field, how much farther have you to go when you have walked 200 m?

3 A garden has a concrete path running down each side of it. If the paths are 50 cm wide, and there are 48 sq m of concrete, what is the length of the garden?

4 What is the total surface area of a cube which has a side of length $4\frac{1}{2}$ cm?

5 A flag measuring 1 m × 1 m is green, except for 1000 sq cm, which is red. What fraction of it is green?

6 A football pitch is laid out in a field 170 m × 140 m. The pitch itself is 102 m × 74 m. What area of the field is left for spectators?

7 If 2440 square stones of side 20 cm are required to pave a court, find the area of the court in sq m.

8 Find the area of the shaded portions of these figures, which are squares of side 16 cm.

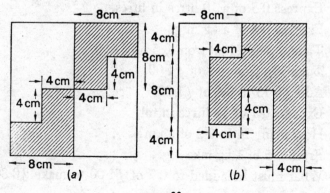

(a)   (b)

DECIMAL NOTATION (1)

First try these

1 Write three-tenths as a decimal.
2 Write seven-tenths as a decimal.
3 What decimal represents eight-tenths?
4 What decimal represents five-tenths?
5 What fraction is represented by 0·7?
6 What fraction is represented by 0·9?
7 How many cm in 0·4 of a metre?
8 How many m in 0·3 of a kilometre?
9 How many cm in 0·7 of a m?
10 How many m in 0·5 of a km?
11 How many seconds in 0·6 of a minute?
12 How many seconds in 0·4 of a minute?
13 What is 0·2 of an hour in minutes?
14 What is 0·5 of an hour in minutes?
15 How many grammes in 0·9 of a kilogramme?
16 How many pence in £0·80?
17 Change 0·5 litre to ml.
18 How many kg in 0·6 of a tonne?
19 How many kg in 0·7 of a tonne?
20 Express 0·5 of a 10 litres in litres.
21 Express 0·5 of a kg in g.
22 Take 0·3 of £1·00 from 0·5 of £1·00.
23 Take 50p from 0·8 of £1·00.
24 Add £0·15 to 0·6 of £1·00.
25 Express 0·1 of 1¼ litres in ml.
26 How many m in 0·2 of a km?
27 From 0·5 of a kg take 3 g.
28 What must be added to 0·7 of £1·00 to make £0·80?

DECIMAL NOTATION (1)

Now try these

1 Write eleven-tenths and nineteen-tenths as decimals.
2 Write as mixed numbers: 2·7, 3·6, 10·2, 13·1, 15·9.
3 Express as decimals: $\frac{13}{10}$, $\frac{4}{10}$, $\frac{1}{5}$, $\frac{14}{20}$, $\frac{36}{40}$.
4 Express in cm 0·2 of 3 m.
5 What is the difference between 0·5 of a m and 0·5 of 2 m?
6 Write 0·9 of a km in terms of m.
7 What is 1·4 km, 2·3 km, 7·9 km. (Answer in metres.)
8 Express in kilogrammes using decimal form: 400 g, 700 g, 1500 g, 2600 g.
9 How many pence in 0·5 of £1·50?
10 Add 0·8 of 50 p to 0·3 of 30p.
11 Write in kg: 5·3 tonnes, 8·7 tonnes, 14·4 tonnes.
12 How many kg in 0·1 of a tonne, 0·5 of a tonne 1·5 of a tonne?
13 Write in decimal form: $11\frac{4}{20}$ tonnes, $5\frac{3}{10}$ g, $6\frac{9}{10}$ kg.
14 Express 5 kg as a decimal fraction of 50 000 g.
15 How many times can 0·5 litre be taken from 2 litres?
16 To 0·8 of 20 litres add 500 ml.
17 Express 1·2 min, 3·7 min and 0·8 min as seconds.
18 How many hours in 0·5 of a week?
19 Take 1·9 h from 3·7 h. (Answer in h and min.)
20 Find the difference between 0·4 of £1·50 and 0·9 of £5·00.
21 Write in m: 0·6 km, 1·9 km, 0·2 km, 1·1 km.
22 Add 0·5 times a metre to 6·5 times 1 cm.
23 How many g in 0·8 times 3 kg?
24 How many litres in 0·3 times $12\frac{1}{2}$ litres?
25 Express as decimals: $\frac{19}{10}$, $\frac{46}{20}$, $\frac{68}{40}$, $\frac{96}{80}$.
26 To 0·5 of 1 kg add 0·5 of 3 kg.
27 What is the value in cm of 0·1 of 6 metres?
28 Express 7500 ml as a decimal fraction of 30 litres.

DECIMAL NOTATION (2)

All can try these

1 Write as vulgar fractions: 0·2, 0·03. (Answer in lowest terms.)

2 Express in decimal form: $\frac{11}{100}$, $\frac{29}{100}$.

3 Write first as hundredths, and then as decimals: $\frac{1}{4}$ and $\frac{3}{4}$.

4 Express as vulgar fractions in their lowest terms: 0·16, 0·84.

5 Change into decimal form: $\frac{9}{50}$, $\frac{71}{100}$.

6 What are the values of 0·25 and 0·75 of 10 p?

7 Find the difference between £1·50 and 0·25 of £5·00.

8 What decimal fraction of 75 p is 15 p?

9 Write as decimals: $1\frac{13}{20}$, $5\frac{39}{50}$.

10 Express as mixed numbers (in their lowest terms): 1·05, 2·95.

11 Find 0·25 of 1 tonne (answer in kg), and 0·75 of 1 kg. (Answer in g.)

12 What decimal fraction of a tonne is 170 kg?

13 Write 2 kg 2 g as a decimal fraction of 8 kg 8 g.

14 How many cm in 0·41 of 1 m?

15 Add 10p to 0·32 of £10·00. (Answer in £.)

16 How many m must be added to 0·75 km to make 2 km?

17 If 1p is 0·01 of a boy's money, how much has he?

18 Find the difference between 0·25 of 80 litres, and 15 litres. (Answer in ml.)

19 How many litres in 0·07 of 200 litres?

20 How many seconds in 0·75 of 2 h?

21 What decimal fraction of 14 weeks 2 days is 19 days?

22 Write $10\frac{3}{100}$ and $\frac{98}{100}$ in decimal notation.

23 Write as hundredths, and then as decimals: $\frac{23}{25}$, $\frac{19}{20}$, $\frac{37}{50}$.

24 What is the value of 0·65 of 20 litres. (Answer in litres.)

DECIMAL NOTATION (3)

All can try these

1 What is the value of 7 in each of the following numbers: 70, 7, 0·7, 0·07?

2 Write in decimal form: $\frac{17}{100}$, $\frac{23}{100}$, $\frac{49}{100}$, $\frac{75}{100}$, $\frac{86}{100}$.

3 What is the value of 3 in each of the following numbers: 3, 0·3, 0·03, 0·003?

4 Express $\frac{1}{8}$ in decimal notation.

5 Express as vulgar fractions in their lowest form: 0·375, 0·625, 0·875.

6 Write as hundredths, and then as decimals: $\frac{38}{50}$, $\frac{33}{25}$, $\frac{47}{20}$.

7 Write as thousandths, and then as decimals: $\frac{1}{500}$, $\frac{21}{200}$, $\frac{168}{250}$.

8 Express as decimals: $2\frac{9}{1000}$, $15\frac{3}{100}$, $\frac{986}{1000}$, $\frac{1592}{1000}$.

9 Express as vulgar fractions in their lowest terms: 0·068, 0·004, 0·425.

10 Write as vulgar fractions in their lowest terms: 0·008, 0·775, 0·103.

11 How much greater is 300 kg than 0·125 tonne? (Answer in kg.)

12 What is the value in metres of 0·281 km, 0·674 km?

13 How far is 0·125 km, 0·75 km, 0·875 km? (Answer in m.)

14 What weight is contained in 0·375 tonne (answer in kg), and 0·125 kg (answer in g)?

15 How many litres in 0·875 of 16 litres?

16 Find how many minutes there are in 0·25 h, 1·55 h, 2·625 h.

17 If 0·005 of a man's money is £1, how much has he?

18 Write as decimals: $\frac{3}{8}$, $\frac{13}{40}$, $\frac{299}{500}$, $\frac{16}{20}$, $\frac{3}{10}$, $\frac{2768}{1000}$.

19 Subtract 0·15 of 10 kg from 0·55 of 8 kg.

20 Express $1\frac{1}{2}$p as a decimal fraction of 10p, and 65p as a decimal fraction of £1·00.

21 Write as decimals: $4\frac{19}{1000}$, $6\frac{8}{100}$, $3\frac{217}{1000}$, $20\frac{3}{10}$.

22 Express 9 kg as a decimal fraction of 180 kg.

DECIMALS ADDITION

First try these

Set 1

1	0·4	2	0·9	3	4·2	4	2·9
	0·7		1·5		0·8		0·6
	0·8		2·3		3·5		1·3
	+0·3		+0·1		+0·4		+3·6

5	0·03	6	0·31	7	0·26	8	1·21
	0·67		0·56		0·15		0·39
	0·70		0·48		0·09		5·50
	+0·91		+0·79		+0·84		+3·08

9	7·65	10	0·10	11	9·49	12	6·84
	4·21		2·58		5·32		7·20
	0·07		6·07		1·84		3·17
	+0·42		+0·01		+4·91		+0·03

Set 2

1	12·97	2	15·80	3	16·45	4	19·57
	3·25		0·58		1·98		34·08
	8·64		11·36		24·73		26·44
	+10·52		+20·14		+18·82		+31·02

5	50·39	6	28·23	7	33·62	8	45·60
	43·02		65·70		79·85		19·83
	0·89		91·34		12·06		58·08
	+31·76		+13·50		+5·37		+27·34

9	103·50	10	98·56	11	125·91	12	88·35
	28·21		35·07		47·63		46·93
	21·37		16·48		10·08		30·12
	+0·59		+180·90		+0·95		+1·80

DECIMALS ADDITION

Now try these

Set 1

1		2		3		4	
	1·9		0·42		0·57		1·90
	4·6		0·31		0·49		3·25
	0·5		0·85		0·03		0·41
	+2·7		+0·76		+0·38		+2·84

5		6		7		8	
	0·301		0·852		0·236		1·432
	0·125		0·197		0·912		2·075
	0·619		0·400		0·835		0·643
	+0·784		+0·015		+0·386		+3·197

9		10		11		12	
	3·210		12·863		36·071		42·193
	8·975		4·109		2·709		28·066
	1·083		15·326		48·865		57·587
	+5·191		+27·014		+19·546		+64·185

Set 2

1		2		3		4	
	18·7		14·72		0·93		91·03
	39·5		6·85		13·08		14·79
	42·0		50·13		26·58		59·32
	+25·6		+19·16		+54·61		+87·95

5		6		7		8	
	7·57		12·03		286·34		187·51
	29·05		196·54		129·86		402·9
	103·62		96·6		0·3		61·74
	+68·40		+ 2·75		+10·09		+15·2

9		10		11		12	
	152·917		361·072		515·017		273·815
	38·456		53·813		83·21		532·198
	205·129		20·105		0·9		700·51
	+94·732		+194·693		+672·38		+10·7

DECIMALS SUBTRACTION

First try these

Set 1

1	0·7 − 0·2	2	1·8 − 0·5	3	3·2 − 2·7	4	4·6 − 2·9
5	8·6 − 5·3	6	9·1 − 7·6	7	10·0 − 4·9	8	14·5 − 10·8
9	0·49 − 0·27	10	0·71 − 0·36	11	0·58 − 0·07	12	0·32 − 0·16
13	4·03 − 2·91	14	6·85 − 4·93	15	12·72 − 8·06	16	10·05 − 3·59
17	17·62 − 15·38	18	25·08 − 19·53	19	63·74 − 54·06	20	83·29 − 47·62
21	108·70 − 59·41	22	136·25 − 98·65	23	174·09 − 68·53	24	200·00 − 19·76

Set 2

1	1·6 − 0·8	2	1·25 − 0·73
3	9·5 − 2·4	4	13·2 − 11·6
5	10·85 − 7·49	6	9·90 − 1·07
7	67 − 39·8	8	5·36 − 1·28
9	34·02 − 9·11	10	100 − 15·62
11	0·48 − 0·09	12	1·05 − 0·63
13	123·6 − 85·7	14	73·84 − 34·95
15	138·17 − 76·53	16	201·24 − 68·59

DECIMALS SUBTRACTION

Now try these

Set 1

1   4·1
  – 2·7
	2 8·5
	  – 6·2
		3 9·3
		  – 7·8
			4 13·2
			  – 9·4

5 16·45 – 2·97	**6** 25·03 – 18·76	**7** 41·82 – 28·03	**8** 93·45 – 49·57

9   0·406
  – 0·139
	10 0·971
	  – 0·508
		11 5·376
		  – 2·649
			12 12·073
			 – 10·814

13 73·210 – 15·308	**14** 90·519 – 68·453	**15** 100·000 – 47·609	**16** 126·813 – 71·084
17 251·728 – 139·642	**18** 347·253 – 198·501	**19** 638·025 – 259·418	**20** 711·028 – 586·139
21 417·429 – 32·534	**22** 312·576 – 291·915	**23** 930·532 – 846·095	**24** 867·541 – 99·738

Set 2

1 10·2 – 1·9   **2** 0·75 – 0·29

3 20·01 – 3·6   **4** 12·3 – 7·8

5 34 – 0·34   **6** 48·206 – 15·817

7 96·725 – 89·315   **8** 123 – 0·123

9 72·380 – 24·642   **10** 14·345 – 8·629

11 253·05 – 84·218   **12** 529 – 0·76

13 854·264 – 639·182   **14** 369·015 – 229·824

15 999·385 – 567·949   **16** 1000 – 31·267

DECIMALS PROBLEMS (1)

First try these

1 Town A had 0·45 cm of rain in one day and Town B had 0·20 cm. Find (a) which place had more rain, and (b) how much more rain this was.

2 What is the difference in speeds between 89·23 km/h and 103·17 km/h?

3 A rectangular garden is 97·125 m long and 35·75 m wide. How many m of fencing are required altogether for one long and one short side?

4 Which is the greater, and by how much: (a) 1·75 of 16p or (b) 1·25 times 24p?

5 In a motor race the average speed of the winning car was 192·68 km/h and of the second car 179·57 km/h. How much faster was the winning car?

6 Add, in decimal form: $17\frac{1}{2}$, $16\frac{7}{8}$, $\frac{1}{20}$ and $1\frac{3}{5}$.

7 Subtract the difference between 8·6 and 0·86 from the sum of 15·2 and 1·52.

8 Find the value of 3·625 km + 5·25 km + 6·05 km. Give the answer in metres.

9 A man wrote down 52·5 litres instead of 5·25 litres. What difference did this make? Give the answer (a) in litres and a decimal of a litre, and (b) in ml.

10 Work in decimal form: $5\frac{1}{4} - 1\frac{1}{8} + 11\frac{2}{5} - 3\frac{17}{20}$.

11 Express 1 min 36 sec as a decimal of 16 min.

12 In four days a grocer sold 16·5 kg, 17·375 kg, 14·625 kg and 19·875 kg of sweets. What total weight of sweets did he sell in the 4 days? Give your answer (a) in decimal form, and (b) in kg and g.

13 How many halfpence are there in (a) £0·12$\frac{1}{2}$, (b) £0·20$\frac{1}{2}$.

14 Find the difference between 6892 hundredths and 637 tenths. (Answer in units.)

15 Express 0·6 of 30 cm as a decimal of 1 m.

16 What must be added to the sum of 87·2 and 19·68 to make the result equal to 110·03?

DECIMALS PROBLEMS (1)

Now try these

1 Find the sum of 3·75 m, 7·93 m, 8·2 m and 10·01 m.

2 A tank held 84·375 litres of water. When 29·625 litres had been drawn off, how much remained? Give the answer (*a*) in decimal form, and (*b*) in litres and ml.

3 Find the value of 1·875 tonne + 2·25 tonne + 1·5 tonne. Give the answer in kg.

4 An express train took the following times over successive quarter km: 15·4, 15·5, 15·4, 15·5, 15·9, 16·6, 17·3, 18·4 sec. How long did it take to cover the 2 km? (Answer in min and sec.)

5 Find the difference, in m, between 6·125 km and 5061 m. (Give the answer in metres.)

6 Work in decimal form: (*a*) $19\frac{1}{5} - 13\frac{7}{8}$, (*b*) $101\frac{3}{10} - 89\frac{3}{4}$.

7 A housewife owes £13·375, £8·25 and £12·65. How much does she owe altogether?

8 In a week, a town had the following amounts of sunshine: 12·8 h, 9·3 h, 4·7 h, 10·2 h, 5·9 h, 8·6 h, 11·1 h. How many hours of sunshine did it have during that week?

9 A cargo ship steamed 480·37 sea miles on Monday, but had to reduce speed the following day because of rough seas, and covered only 340·59 sea miles. How much farther did it travel on Monday than on Tuesday?

10 Find the value of 0·33 of 50p, and express the result as a decimal fraction of 22p.

11 Find the sum of 8·75, 9·125, 3·25, 4·875, 2·9, 3·5, 1·6, 6.

12 Add in decimal notation: $\frac{7}{50}$, $\frac{16}{25}$, $\frac{9}{20}$, $\frac{3}{10}$, $\frac{5}{8}$, $\frac{4}{5}$, $\frac{3}{4}$, $\frac{1}{2}$.

13 (*a*) 112·85 tonnes – 4·25 tonnes. (Answer in kg.)
(*b*) 25·5 kg – 11·625 kg. (Answer in g.)

14 Find the value of 9·085 – 17·036 + 22·8 – 11·76 + 0·053 + 16 – 1·008.

15 Subtract 0·025 of 1 km from 0·375 of 200 m. (Answer in m.)

DECIMALS MULTIPLICATION

First try these

Set 1

1 $1 \cdot 2 \times 10$	**2** $0 \cdot 5 \times 10$	**3** $2 \cdot 7 \times 10$
4 $0 \cdot 9 \times 10$	**5** $3 \cdot 4 \times 2$	**6** $5 \cdot 8 \times 3$
7 $1 \cdot 6 \times 5$	**8** $6 \cdot 3 \times 7$	**9** $0 \cdot 8 \times 9$
10 $1 \cdot 9 \times 10$	**11** $7 \cdot 4 \times 11$	**12** $0 \cdot 6 \times 13$
13 $3 \cdot 6 \times 8$	**14** $9 \cdot 2 \times 12$	**15** $8 \cdot 1 \times 10$
16 $12 \cdot 4 \times 10$	**17** $7 \cdot 3 \times 20$	**18** $0 \cdot 4 \times 100$
19 $2 \cdot 9 \times 100$	**20** $13 \cdot 6 \times 100$	**21** $1 \cdot 8 \times 2 \cdot 1$
22 $0 \cdot 3 \times 4 \cdot 5$	**23** $1 \cdot 2 \times 10 \cdot 7$	**24** $3 \cdot 9 \times 0 \cdot 1$
25 $4 \cdot 6 \times 0 \cdot 7$	**26** $20 \cdot 7 \times 6$	**27** $13 \cdot 1 \times 2 \cdot 5$
28 $14 \cdot 8 \times 1 \cdot 7$	**29** $6 \cdot 3 \times 2 \cdot 1$	**30** $5 \cdot 7 \times 50$
31 $25 \times 0 \cdot 4$	**32** $12 \cdot 5 \times 0 \cdot 2$	

Set 2

1 $0 \cdot 61 \times 10$	**2** $0 \cdot 05 \times 10$	**3** $2 \cdot 31 \times 10$
4 $4 \cdot 08 \times 10$	**5** $1 \cdot 45 \times 6$	**6** $3 \cdot 14 \times 2$
7 $5 \cdot 72 \times 3$	**8** $0 \cdot 53 \times 10$	**9** $0 \cdot 01 \times 100$
10 $4 \cdot 27 \times 100$	**11** $6 \cdot 19 \times 10$	**12** $8 \cdot 34 \times 20$
13 $0 \cdot 087 \times 10$	**14** $0 \cdot 003 \times 10$	**15** $0 \cdot 045 \times 100$
16 $1 \cdot 276 \times 10$	**17** $9 \cdot 3 \times 2 \cdot 14$	**18** $4 \cdot 81 \times 3 \cdot 5$
19 $2 \cdot 01 \times 8 \cdot 6$	**20** $12 \cdot 3 \times 0 \cdot 06$	**21** $0 \cdot 57 \times 0 \cdot 5$
22 $0 \cdot 32 \times 0 \cdot 9$	**23** $0 \cdot 75 \times 8 \cdot 4$	**24** $4 \cdot 3 \times 0 \cdot 7$
25 $10 \cdot 23 \times 5 \cdot 1$	**26** $2 \cdot 8 \times 6 \cdot 7$	**27** $1 \cdot 29 \times 0 \cdot 4$
28 $0 \cdot 34 \times 27$	**29** $11 \cdot 23 \times 10$	**30** $12 \cdot 97 \times 0 \cdot 3$
31 $0 \cdot 61 \times 20 \cdot 5$	**32** $0 \cdot 008 \times 100$	

DECIMALS MULTIPLICATION

Now try these

Set 1

1 $19 \cdot 7 \times 10$

2 $0 \cdot 34 \times 100$

3 $1 \cdot 246 \times 1000$

4 $0 \cdot 004 \times 10$

5 $10 \cdot 035 \times 100$

6 $0 \cdot 003 \times 1000$

7 $24 \cdot 852 \times 10$

8 $47 \cdot 198 \times 100$

9 $1 \cdot 8 \times 0 \cdot 6$

10 $2 \cdot 5 \times 2 \cdot 5$

11 $14 \cdot 2 \times 13$

12 $54 \cdot 3 \times 12 \cdot 1$

13 $0 \cdot 95 \times 16 \cdot 8$

14 $28 \times 0 \cdot 009$

15 $37 \times 5 \cdot 63$

16 $0 \cdot 293 \times 45$

17 $2 \cdot 565 \times 19$

18 $427 \times 0 \cdot 004$

19 $70 \cdot 3 \times 1 \cdot 2$

20 $6 \cdot 03 \times 11$

21 $7 \cdot 5 \times 1 \cdot 8$

22 $0 \cdot 81 \times 17$

23 $100 \cdot 1 \times 4 \cdot 6$

24 $39 \cdot 65 \times 0 \cdot 9$

25 $0 \cdot 607 \times 54$

26 $115 \cdot 9 \times 8 \cdot 3$

27 $0 \cdot 72 \times 49 \cdot 7$

28 $7 \cdot 528 \times 65$

29 $25 \cdot 6 \times 10 \cdot 03$

30 $1 \cdot 37 \times 20 \cdot 9$

31 $508 \cdot 3 \times 0 \cdot 07$

32 $59 \cdot 4 \times 2 \cdot 48$

Set 2

1 $6 \cdot 002 \times 11$

2 $0 \cdot 015 \times 38$

3 $53 \cdot 6 \times 9 \cdot 7$

4 $0 \cdot 9 \times 30 \cdot 06$

5 $0 \cdot 112 \times 98$

6 $15 \cdot 6 \times 23$

7 $48 \cdot 25 \times 16 \cdot 4$

8 $50 \cdot 4 \times 8 \cdot 07$

9 $2 \cdot 158 \times 102$

10 $3 \cdot 81 \times 62 \cdot 3$

11 $0 \cdot 72 \times 14 \cdot 9$

12 $13 \times 0 \cdot 59$

13 $52 \cdot 6 \times 0 \cdot 18$

14 $93 \cdot 1 \times 50 \cdot 7$

15 $46 \cdot 8 \times 7 \cdot 36$

16 $117 \cdot 5 \times 6 \cdot 4$

17 $126 \times 0 \cdot 005$

18 $41 \times 0 \cdot 906$

19 $0 \cdot 371 \times 75$

20 $8 \cdot 926 \times 70$

21 $100 \cdot 5 \times 10 \cdot 04$

22 $76 \times 81 \cdot 5$

23 $93 \cdot 4 \times 0 \cdot 19$

24 $49 \times 0 \cdot 006$

25 $60 \cdot 3 \times 50$

26 $48 \times 20 \cdot 004$

27 $0 \cdot 625 \times 40$

28 $13 \cdot 21 \times 10 \cdot 5$

29 $432 \times 0 \cdot 016$

30 $0 \cdot 57 \times 0 \cdot 3$

31 $73 \cdot 9 \times 21 \cdot 6$

32 $96 \times 68 \cdot 4$

DECIMALS DIVISION

First try these

Set 1

1 429÷10	**2** 532÷100	**3** 94÷10
4 37÷100	**5** 12÷8	**6** 67÷5
7 45÷6	**8** 27÷12	**9** 7·2÷9
10 12·1÷11	**11** 94·2÷3	**12** 7÷10
13 81÷12	**14** 10·7÷2	**15** 86÷20
16 63·27÷9	**17** 2·5÷10	**18** 0·48÷6
19 1·65÷11	**20** 13·8÷100	**21** 0·9÷10
22 7·04÷20	**23** 0·54÷9	**24** 0·372÷12
25 0·517÷11	**26** 63÷0·3	**27** 6·6÷0·6
28 10÷0·5	**29** 8÷0·4	**30** 10·5÷0·7
31 2·7÷0·9	**32** 14÷0·8	

Set 2

1 57÷100	**2** 889÷10	**3** 1·6÷10
4 87÷30	**5** 38·5÷7	**6** 7·68÷12
7 0·96÷8	**8** 28·8÷9	**9** 45·1÷11
10 0·75÷5	**11** 13·08÷12	**12** 5·13÷9
13 2·6÷1·3	**14** 6·3÷1·4	**15** 7·7÷2·2
16 13·52÷5·2	**17** 16·81÷4·1	**18** 6·25÷2·5
19 23·28÷8	**20** 25·26÷6	**21** 33·55÷11
22 2·808÷2·6	**23** 4·065÷1·5	**24** 5·439÷3·7
25 6·08÷1·6	**26** 37·62÷9	**27** 12·978÷4·2

DECIMALS DIVISION

Now try these

Set 1

1 $58 \div 100$	**2** $2 \cdot 74 \div 10$	**3** $9 \div 1000$
4 $5 \cdot 6 \div 4$	**5** $93 \div 12$	**6** $66 \div 8$
7 $28 \cdot 6 \div 11$	**8** $9 \cdot 72 \div 9$	**9** $0 \cdot 378 \div 6$
10 $0 \cdot 984 \div 12$	**11** $2 \cdot 92 \div 4$	**12** $5 \cdot 6 \div 0 \cdot 7$
13 $6 \div 1 \cdot 25$	**14** $0 \cdot 38 \div 0 \cdot 5$	**15** $2 \cdot 8 \div 0 \cdot 25$
16 $1 \cdot 44 \div 7 \cdot 2$	**17** $6 \cdot 82 \div 1 \cdot 1$	**18** $0 \cdot 252 \div 2 \cdot 1$
19 $7 \cdot 2 \div 0 \cdot 008$	**20** $2 \cdot 565 \div 19$	**21** $100 \div 0 \cdot 625$
22 $0 \cdot 48 \div 0 \cdot 75$	**23** $712 \cdot 8 \div 132$	**24** $78 \cdot 045 \div 12 \cdot 9$
25 $2 \cdot 38 \div 0 \cdot 007$	**26** $0 \cdot 034 \div 6 \cdot 8$	**27** $116 \cdot 58 \div 8 \cdot 7$
28 $156 \cdot 25 \div 12 \cdot 5$	**29** $259 \cdot 2 \div 3 \cdot 6$	**30** $0 \cdot 24 \div 0 \cdot 25$
31 $58 \cdot 48 \div 13 \cdot 6$	**32** $1 \cdot 125 \div 0 \cdot 375$	

Set 2

1 $3 \div 2 \cdot 5$	**2** $0 \cdot 9 \div 100$	**3** $0 \cdot 096 \div 4 \cdot 8$
4 $107 \cdot 1 \div 21$	**5** $42 \div 0 \cdot 03$	**6** $4 \div 1000$
7 $13 \cdot 75 \div 0 \cdot 625$	**8** $6 \cdot 664 \div 0 \cdot 34$	**9** $75 \cdot 6 \div 180$
10 $88 \cdot 83 \div 30$	**11** $124 \div 62 \cdot 5$	**12** $0 \cdot 918 \div 0 \cdot 017$
13 $142 \cdot 2 \div 7 \cdot 2$	**14** $210 \div 0 \cdot 35$	**15** $16 \cdot 8 \div 0 \cdot 042$
16 $36 \cdot 801 \div 4 \cdot 23$	**17** $6 \cdot 08 \div 0 \cdot 475$	**18** $23 \cdot 37 \div 24 \cdot 6$
19 $1827 \div 900$	**20** $0 \cdot 286 \div 0 \cdot 44$	**21** $45 \cdot 847 \div 12 \cdot 7$
22 $136 \cdot 74 \div 8 \cdot 6$	**23** $0 \cdot 943 \div 0 \cdot 023$	**24** $4 \cdot 958 \div 134$
25 $201 \cdot 64 \div 14 \cdot 2$	**26** $32 \cdot 83 \div 93 \cdot 8$	**27** $0 \cdot 252 \div 28$

DECIMALS PROBLEMS (2)

First try these

1 A school ordered 50 litres of milk, but the dairyman sent 1·4 times as much. How many litres did the school actually receive?

2 How many revolutions does a wheel, with circumference of 4·2 m, make in covering a distance of 107·52 m?

3 How far does a train travel in 3·75 h, if it keeps up a steady speed of 70 km/h? (Answer in km.)

4 How many paving-stones were used to make a curb 200 m long, if ¼ of the length was made up of stones 0·5 m long and the remaining length with stones 0·75 m long?

5 Divide 2·89 by 1·7, and multiply your result by 1·5.

6 How many strips of paper, each 1·2 cm wide, can be cut from a sheet 20 cm wide? What width of the sheet will be left over?

7 A distance is measured in cm. Take ⅜ of it and the answer is 3 m. What is the distance in cm?

8 What weight of sand is there in 3·5 loads, if each load contains 4·75 tonnes? Give the answer (a) in decimal form, and (b) in kg.

9 The rim of a wheel is always 3·14 times the diameter of the wheel. Find the distance round the rim if the diameter is (a) 68 cm (answer in cm); (b) 15 m (answer in m).

10 The speedometer shows that a car is moving at 69·6 km/h. How far short of a speed of 100 km/h is this?

11 Workmen built a wall 516·25 m long in 12·5 days. How many m of wall did they build per day?

12 How far does a man walk in 6·4 h, at a steady rate of 5·125 km/h?

13 Divide 96·88 by 2·8, and multiply the result by 0·5.

14 A sheet of paper 50·2 sq cm in area is divided into 5 equal parts. Find the area of each part.

15 A bicycle was priced at £31·75, but during a sale it was sold at 0·8 times this. What was the selling price?

DECIMALS PROBLEMS (2)

Now try these

1 A liner keeps up a steady speed of 29·5 sea miles per h for 11 h. How many sea miles does it cover in that time?

2 Find the area of a rectangle 9·3 cm long and 4·15 cm wide.

3 Multiply 4·5 by itself, then divide the result by 27.

4 An express train travels at an average speed of 84·7 km/h for 3·65 h. What distance does it travel in this time?

5 Give the value of 0·625 of 1 m as a decimal of 50 cm.

6 How many times can 0·025 be subtracted from 4·175?

7 A parcel weighs 9·375 kg. By how much is its weight less than 10 000 g.

8 A man sets out to walk 42 km. He walks for 4·25 h at 6·12 km/h. (*a*) How far has he still to go? (*b*) At what speed must he walk to complete the remainder of the journey in exactly 3 h?

9 Find the cost of 12 articles at £0·875 each.

10 The smaller of two numbers is 39·61, and their sum is 85·01. Find the product of the two numbers.

11 A saw cuts through a piece of steel, 140·7 cm thick, at a steady speed of 8·04 cm per min. How many minutes and seconds does it take?

12 Add together 4·7, 2·3, 6·4, 8·5, 9·2 and 7·3, and multiply the result by the largest of these numbers.

13 A car travels at a steady speed of 63·5 km/h. How many hours and minutes will it take to travel 298·45 km?

14 If 508·60 American dollars are distributed equally among 20 persons, how much will each get?

15 Multiply 78 by 0·78, and divide your result by 5·2.

16 What quantity is needed to give 8·75 ml to each of 250 persons. (Answer in ml.)

17 Express 35p as a decimal of £1·00. Then find the value of 500 × 35p.

GENERAL REVISION

First try these

Add

1
   29 301
   72 859
      625
 + 14 004

2

km	m
2	506
19	714
3	41
+ 11	100

3 $1\frac{2}{5} + 4\frac{3}{4}$

Subtract

4

tonnes	kg
59	206
− 7	491

5

kg	g
12	30
− 5	200

6
  124·50
 − 79·68

Multiply

7
  508
 × 27

8

m	cm
15	82
	× 2

9 £
 13·59½
 × 10

Divide

£

10 $99·49\frac{1}{2} \div 9$

11 $1·65 \div 0·33$

12

kg	g
40	800 ÷ 12

13 Write in figures five hundred thousand, two hundred and six.

14 Using a scale of 1 cm to 5 m, draw the following figures: (*a*) a rectangle 15 m × 10 m; (*b*) a triangle which has each of its sides 8·5 m.

15 If 3 times a number is 10 more than four score, find (*a*) the number, and (*b*) 3 times half the number.

16 How many articles each costing £0·36 can be bought for £3·96?

17 A litre of water weighs 1 kg. Find the weight of 984 ml. (Answer in g.)

18 A question in an arithmetic book read, "Add $\frac{1}{2} + \frac{1}{4} + \frac{1}{6}$." The answer given was $\frac{11}{12}$. Find the missing number.

19 A man spends 0·15 of his income on rent and 0·6 on household expenses. He has £800·00 a year left. What is his income?

GENERAL REVISION

Now try these

Add

1	£	2	litres	ml		3	$\frac{4}{15} + \frac{1}{3} + \frac{2}{5}$
	216·08		38	215			
	197·53½		16	476			
	0·48½		34	19			
	+ 309·00		+ 27	117			

Subtract

4	673·015	5	days	h	min	6	km	m
	− 84·209		36	15	48		91	70
			− 19	21	29		− 35	450

Multiply

7	$4\frac{2}{3} \times 1\frac{2}{7} \times 3\frac{1}{5}$	8	23·6	9	km	m
			× 1·09		15	90
						× 2

Divide

	£					h	min	sec
10	200·88 ÷ 8	**11**	0·945 ÷ 0·27	**12**		53	28	0 ÷ 2

13 Write in figures one million, six thousand and ninety-five.

14 A book contained 36 608 words. If there were 26 lines on each page, and 11 words on each line, how many pages were there?

15 A tank held 160 litres. After 50 litres had been drawn off, what was the remainder worth at 24p per litre?

16 A map is drawn to a scale of $\frac{1}{4}$ cm to the km. What area of land is represented by a map $8\frac{1}{4}$ cm × $5\frac{1}{2}$ cm?

17 A man smokes 4 g of tobacco daily. How much does he smoke in (*a*) January, (*b*) from 1st January to 31st December 1976, inclusive? (Answers in g.)

18 How many halfpence are there in £6·78?

19 A man buys 80 doz. eggs at £0·35 a doz. and finds $\frac{1}{40}$ are broken and $\frac{1}{80}$ bad. He sells the remainder at £0·45 a dozen. How much profit does he make?

MISSING FIGURES

All can try these

Copy these examples, and fill in the missing figures, or replace the letters

1		2		3		4	
	2*		49		*26		13*6
	19		67		21*		7
	*3		30		398		*39
	+35		+**		+2*9		+*70*
	131		201		948		7474

5		6		7		8	
	291		***		*2*8		10**9
	−1**		−228		−3*7*		−*375
	165		207		1242		188*

9		10		11		12	
	1		***		13A		ABB
	×6		×7		×A		×B
	1302		2723		67A		3BB

13		14		15	16	
	A2B		10A2	3)24*6	A)36A5
	×B		×A	*02	525
	11AB		A168		

```
         A1                    A2                   211
17  1A)40A          18  2A)102A          19  AA)92B5
         A9                    9B                   BB
         1A                    BA                   AB
         1A                    A8                   AA
                               1B                   A5
                                                    AA
                                                     1
```

UNEQUAL SHARING (1)

All can try these

1 Divide £1·00 into two parts, one 10p more than the other.

2 Divide 50p into two parts, one 6p more than the other.

3 A piece of wood 3 m long is cut into two pieces, so that one piece is 6 cm longer than the other. What are the lengths of the pieces?

4 Divide £10·00 into two parts, so that one is £2·50 more than the other.

5 Two parcels together weigh 18 kg. One weighs 4 kg more than the other. Find the weight of each.

6 In a school of 480 children there are 32 more girls than boys. How many boys and how many girls are there?

7 From A to B is $10\frac{1}{2}$ km. One man starts out from A and another from B, and they walk towards each other. They meet when the man from A has walked $1\frac{1}{2}$ km more than the man from B. How many km from A is their meeting-place?

8 A duty lasting 2 h is divided between 2 men, so that one is on duty 20 min longer than the other. How long is each man's turn of duty?

9 An 8 m pole is painted in two colours, red and white. If the portion painted red is 2 m longer than the portion painted white, what length of pole is red?

10 A cask holding 25 litres of water was emptied in 2 days. If 3 litres more were drawn off on the second day than on the first, find how much was used on the second day.

11 Divide the period between 08:24 and 11:38 into two periods so that the first is 18 min longer than the second.

12 The distance round the edge of a rectangular field is 280 m. If the length is 20 m greater than the width, find the length and breadth of the field.

13 A and B had similar lunches except that A had coffee costing 8p. The bill came to 90p. What did B's lunch cost?

AVERAGES

First try these

1 Find the average of
   (a) 8, 10, 4, 2
   (b) 1, 3, 8, 12, 6
   (c) 7p, 3p, 5p
   (d) 10p, 50p, 80p, 20p, 15p
   (e) 7·4, 8·5, 1·3, 2·6
   (f) 0·72, 0·35, 0·19
   (g) £2·50, £3·00, £4·75, £2·25

2 Find the total cost of
   (a) 4 books, if the average cost is 21p each
   (b) 3 toys, if the average cost is 38p each
   (c) 6 articles, if the average cost is 53p each
   (d) 10 kg of apples, if the average cost is 10p per $\frac{1}{2}$ kg

3 In 7 innings a cricketer scored 196 runs. What was his average score?

4 A car used $9\frac{1}{2}$ litres of petrol in travelling 266 km. What was the average distance travelled per litre?

5 The average of 3 numbers is 8. If one is 6 and another 10, what is the third number?

6 Betty is 8 years 2 months, Mary 9 years 10 months and Susan 10 years 9 months. What is their average age?

7 A lorry travels on the average, 20 km to 4 litres of diesel oil. How many litres will it take to travel 290 km?

8 A father and his 3 sons earn £42·00, £30·20, £22·00 and £33·00 per week. What is their average wage?

9 Add all the numbers from 1 to 10 inclusive and find their average. (Express as a decimal.)

10 Three boys ran a race. One took 2 min 7 sec, another 2 min 24 sec and a third took 1 min 59 sec. What was their average time?

AVERAGES

Now try these

1 Find the average of
   (a) 26, 8, 13, 37
   (b) £3·60, £7·95, £5·40
   (c) 4p, 31p, 48p, 15p, 72p
   (d) 97·9, 13·6, 2·1, 4·5, 11·7, 20·8
   (e) 9 kg, 3·5 kg, 8·7 kg, 2·8 kg

2 In two weeks a class of children used the following numbers of bottles of milk: 32, 35, 29, 29, 31, 33, 30, 28, 31, 32. What was the average number of bottles used daily?

3 A girl gained 71 marks in English, 63 in Arithmetic, 69 in History, 57 in Geography and 60 in Nature Study. What was her average mark in these five subjects?

4 Mother spent £18·75 at the grocer's in 4 weeks. In the first three weeks she spent £4·15 per week on the average. How much did she spent in the fourth week?

5 An express train travelled 484 km in $5\frac{3}{4}$ h. If 15 min of this time was spent stopping at stations, what was the average speed of the train, taking into account only the travelling time?

6 Find the average cost of 8 books at 55p each and 4 books at 70p each.

7 Three boys are 135 cm, 129·6 cm and 140·4 cm in height. Find their average height. When another boy joins them, their average height becomes 134·5 cm. What is the height of the fourth boy?

8 In 7 innings at cricket Mr *A* scores 29, 15, 21, 0, 28, 13, 34 runs. In 5 innings Mr *B* scored 37, 8, 16, 9, 20 runs. By how many runs did Mr *A*'s average exceed that of Mr *B*?

9 A car travels for 3 h at an average speed of 40 km/h, and for 7 h at an average speed of 35 km/h. How far does the car travel in 10 h, and what is the average speed for the whole journey?

MORE AVERAGES

First try these

1 Find the average (*a*) age, (*b*) height and (*c*) number of days at school during the year, of the following children

	Age years	Age months	Height cm	Number of days at school
Peter . .	10	9	136	189
Susan . .	11	4	139·5	178
Henry . .	11	7	128·5	184
John . .	10	11	130	165
Anne . .	11	0	140·5	170
Rosemary .	10	8	132·5	182

2 A milkman delivers milk every day of the week. During July he delivered 2697 litres in a certain road. What was the average number of litres delivered in that road each day?

3 A man's salary started at £2000·00 and increased by £150·00 each year. Find his average salary over the first four years.

4 Find the average of $\frac{1}{6}$, $\frac{1}{4}$, $\frac{1}{3}$, $\frac{1}{2}$.

5 In an arithmetic test given to 30 children 1 scored 0: 3 scored 1: 4 scored 4: 11 scored 5: 5 scored 6: 3 scored 8: 2 scored 9 and 1 scored 10. What was the average score? (Answer in decimals.)

6 What is the value of x, if the average of 29, 42, 13, 20, 35, 4, 0, 18 and x is 19?

7 The average weight of a rowing crew of 9 men is 71 kg. The cox weighs 53 kg. Find the average weight of the rest of the crew.

8 Find the average of 0·274, 3·6, 0·008, 9·92, 1·013.

9 Thirty-five children in a class give, on the average, 5p towards a present for a child in hospital. The teacher offers to give an amount which will make the average given by himself and the children 6p. How much must the teacher give?

MORE AVERAGES

Now try these

1 A merchant sells, on the average, 32 litres of paraffin per day from Monday to Thursday inclusive. On Friday he sells 40 litres and on Saturday 60 litres. What is his average daily sale of paraffin during the six days?

2 For ten innings a cricketer's average was 57·5 runs. After another two innings his average was 61 runs. What was his average for the last two innings?

3 Six tonnes of coal costing £32·00 per tonne were mixed with ten tonnes costing £30·00 per tonne. What was the cost of the mixture per ton?

4 A ship travelled, on the average, 480 sea miles a day for 6 days. On the 7th day it increased speed so that the average distance covered per day for the seven days was 485 sea miles. How far did the ship travel on the 7th day, and what was its average speed in knots on the 7th day? (1 knot = 1 sea mile per h.)

5 In a factory 250 men earn an average of £40·00 per week. A number of women are also employed, whose average earnings are £5·00 less per week than those of the men. Find the number of women employed, if the total wages paid out per week is £14550·00.

6 By how much is the average of 1079, 4235, 3862, 1703, 5180, 2934 greater than the average of 3060, 2191, 128, 499, 2373, 5274, 1000?

7 A train took 6 h to travel 540 km. A car took 4 h longer. By how much was the average speed of the car less than that of the train?

8 It costs £23·10 each person per week, on the average, for a father, mother and two children to stay at a hotel in June. In August, the average cost is £26·50 per person, per week. How much does it cost for the family to take a fortnight's holiday in June? How much cheaper is a June holiday for this family than an August holiday?

VOLUME (1)

First try these

1 Find the volume of rectangular blocks with the following measurements

(a) $4 \text{ m} \times 2 \text{ m} \times 2 \text{ m}$ (b) $6 \text{ m} \times 3 \text{ m} \times 1\frac{1}{2} \text{ m}$

(c) $8 \text{ m} \times 6 \text{ m} \times 2\frac{1}{4} \text{ m}$ (d) $9 \text{ cm} \times 5 \text{ cm} \times 4 \text{ cm}$

(e) $10 \text{ cm} \times 9 \text{ cm} \times 7 \text{ cm}$ (f) $4\frac{1}{2} \text{ cm} \times 2 \text{ cm} \times \frac{1}{2} \text{ cm}$

(g) $15 \text{ cm} \times 12 \text{ cm} \times 3\frac{3}{4} \text{ cm}$ (h) $18 \text{ cm} \times 16 \text{ cm} \times \frac{1}{4} \text{ cm}$

2 Find the volume of the following cubes and rectangular blocks in which the length (L), breadth (B) and height (H) are given

	L	B	H		L	B	H
(a)	4 m	4 m	4 m	(b)	8 m	6 m	1 m
(c)	8 m	$2\frac{1}{2}$ m	2 m	(d)	$1\frac{1}{2}$ m	$1\frac{1}{2}$ m	$1\frac{1}{2}$ m
(e)	5 m	$3\frac{1}{4}$ m	2 m	(f)	10 cm	7 cm	7 cm
(g)	16 cm	$4\frac{1}{2}$ cm	$\frac{1}{4}$ cm	(h)	9 cm	9 cm	9 cm

3 Find the volume of cubes and rectangular blocks with the following measurements

(a) $2 \cdot 4 \text{ m} \times 3 \text{ m} \times 2 \cdot 8 \text{ m}$ (b) $2 \cdot 5 \text{ m} \times 2 \cdot 5 \text{ m} \times 2 \cdot 5 \text{ m}$

(c) $12 \text{ cm} \times 8 \text{ cm} \times 1 \text{ cm}$ (d) $3\frac{3}{4} \text{ m} \times 4 \text{ m} \times \frac{1}{4} \text{ m}$

(e) $12 \text{ cm} \times 12 \text{ cm} \times 12 \text{ cm}$ (f) $4 \cdot 75 \text{ cm} \times 3 \text{ cm} \times 5 \text{ cm}$

(g) $6 \text{ cm} \times 10\frac{1}{2} \text{ cm} \times 4\frac{1}{2} \text{ cm}$ (h) $7 \text{ m} \times 7 \text{ m} \times 7 \text{ m}$

4 How many cubic cm are there in

(a) 1 cu m   (b) $1\frac{1}{2}$ cu m

(c) $2\frac{1}{2}$ cu m   (d) 6 cu m

(e) 8 cu m   (f) $4\frac{1}{4}$ cu m

5 How many cubic m are there in

(a) 1 250 000 cu cm   (b) 3 280 000 cu cm

(c) 950 000 cu cm   (d) 10 000 000 cu cm

(e) 7 330 000 cu cm   (f) 50 000 cu cm

VOLUME (1)

Now try these

1 Find the volume of cubes and rectangular blocks with the following measurements

(a) 8 cm × 8 cm × 8 cm   (b) $2\frac{1}{2}$ m × $1\frac{3}{10}$ m × 1 m
(c) 9 m × 7 m × 5 m   (d) 4·8 cm × 5 cm × 1·5 cm
(e) $4\frac{1}{2}$ cm × $3\frac{1}{2}$ cm × 6 cm   (f) 14 cm × 9 cm × $\frac{1}{2}$ cm
(g) $5\frac{1}{4}$ m × $9\frac{1}{2}$ m × 8 m   (h) 13·25 cm × 8·5 cm × 4 cm

2 Find the volume of the following cubes and rectangular blocks in which the length (L), breadth (B) and height (H) are given.   Give the answer in terms of the largest unit used in each case.

	L	B	H		L	B	H
(a)	1 m	2 cm	6 m	(b)	$1\frac{1}{2}$ m	$1\frac{1}{2}$ m	$\frac{1}{2}$ m
(c)	$7\frac{1}{2}$ cm	4 cm	8 cm	(d)	9·5 cm	9·5 cm	9·5 cm
(e)	18 cm	2 m	$2\frac{1}{2}$ m	(f)	$\frac{1}{2}$ m	$1\frac{1}{2}$ m	18 cm
(g)	8·5 cm	5·75 cm	1 cm	(h)	16 cm	9 cm	$1\frac{1}{4}$ m
(i)	12 m	2 m	10 cm	(j)	4·5 cm	3·75 cm	12 cm

3 How many cubic m are there in
(a) 432 000 cu cm   (b) 2 592 000 cu cm
(c) 6 912 000 cu cm   (d) 17 280 000 cu cm

4 How many cubic cm are there in
(a) 54 cu m   (b) 162 cu m
(c) 270 cu m   (d) $101\frac{1}{4}$ cu m
(e) 675 cu m   (f) $6\frac{3}{4}$ cu m

5 Change $7\frac{3}{4}$ cu m into cu cm, and change 17 cu m into cu cm.

6 Find the number of cu m in 1 857 000 cu cm, and in 90 725 000 cu cm.

7 How many cu cm are there in
(a) 2·5 litres, (b) 3·8 litres, (c) 4·7 litres?

VOLUME PROBLEMS

First try these

Make a rough drawing whenever you can, and put in the measurements

1 A cube has a side of 3 cm. Find (*a*) its volume, and (*b*) the area of one of its faces.

2 Find the volume of a brick 23 cm long, 11·5 cm wide and 7·5 cm deep.

3 What is the weight of a piece of wood 1 m × 15 cm × 10 cm, if 1 cu cm of the wood weighs 10 g?

4 Find the volume of a block 4·5 m long, 2·5 m wide and 2 m high.

5 How many cu m are there in a space measuring 12 m × 3 m × 3 m?

6 Find the volume of a cube whose side is $\frac{1}{4}$ m. Give the answer in cu cm.

7 A block of cement is 1000 cm × 500 cm × 250 cm. Find its volume in cu m.

8 A cube of wood has a side of 12·5 cm. A block of wood has sides 30 cm × 12·5 cm × 6·5 cm. Find (*a*) which is the larger, and (*b*) how many more cu cm of wood there are in the larger piece of wood than in the smaller.

9 Fresh soil to a depth of 15 cm is put on a lawn measuring 5 m × 4 m. What is the volume of the soil?

10 How many bricks 7·5 cm × 5·0 cm × 2·5 cm are required to fill a box 30 cm × 20 cm × 20 cm?

11 A children's sand-pit is 4 m × 3 m and it is covered with sand to a depth of 40 cm. How many cu m of sand does it contain?

12 What is the volume of a rectangular box $1\frac{1}{2}$ m long, 1 m wide and 70 cm deep? (Answer in cu m.)

13 A tank measuring $1\frac{1}{2}$ m × 1 m is one-quarter full when it contains water to a depth of 22 cm. How many cu m of water would it hold when full?

14 A cube of side 10 cm is cut from a block of wood 20 cm × 15 cm × 10 cm. How many cu cm of wood remain?

VOLUME PROBLEMS

Now try these

Make a rough drawing whenever you can, and put in the measurements

1 Find the volume of a block of wood 6·2 cm long, 4·5 cm wide and 1·9 cm high. What is the area of its largest face?

2 How many cu m of sand can be placed in a concrete tank measuring 12 m × 9 m × 4 m?

3 If 1000 litres of water occupy 1 cu m, how many litres does a tank 6 m × 4 m × 1½ m hold?

4 The bottom of a box has an area of 30 sq cm. Find the volume of the box if its height is 4½ cm.

5 A beam of timber is 250 cm long, 30 cm wide and 10 cm thick. It weighs 60 kg. Find the weight of 1 cu cm of the wood.

6 Find the volume of a cellar measuring 2 m long and 2 m wide, which has a ceiling 2¼ m above floor level. (Answer in cu m.)

7 A cubic m of water weighs 1000 kg. What weight of water does a tank measuring 4 m × 3½ m × 2 m hold? (Answer in tonnes.)

8 What is the value of a block of stone 1½ m long, 40 cm broad and 1 m thick, at £1·05 per cu m?

9 How many bricks 24 cm × 12 cm × 7½ cm are required to build a wall 9·6 m long, 4·8 m high and 12 cm thick? (Hint: a brick is laid so that its height is 7½ cm.)

10 From a sheet of metal plate, 2 m × 2 m × ⅛ m, a piece 100 cm × 12 cm × 10 cm is cut. What fraction of the original volume is the amount remaining?

11 A rectangular block of wood 90 cm long, 60 cm wide and 45 cm high has a square hole 30 cm × 30 cm × 45 cm. Find the number of cu cm of wood left in the block.

12 400 cu cm of a block weigh 1 kg. Find the weight of a similar block 1 m × 70 cm × 30 cm. (Answer in kg.)

13 How many 1 cm cubes are required to build a cube of side 1⅕ m?

VOLUME (2)

First try these

Make a rough drawing whenever you can, and put in the measurements

1 Find the height of the cubes or rectangular blocks in the following cases, given the volume, length and breadth

	V	L	B		V	L	B
(a)	27 cu m	3 m	3 m	(b)	36 cu m	6 m	3 m
(c)	24 cu m	4 m	4 m	(d)	112 cu m	7 m	4 m
(e)	125 cu cm	5 cm	5 cm	(f)	180 cu cm	9 cm	8 cm
(g)	72 cu m	12 m	12 m	(h)	770 cu cm	11 cm	10 cm

2 Find the missing measurement, length breadth or height, of the cubes or rectangular blocks in the following cases, given the volume and the length of two sides

	V	L	B		V	L	H
(a)	54 cu m	9 m	3 m	(b)	102 cu cm	17 cm	2 cm
(c)	81 cu m	9 m	4 m	(d)	343 cu cm	7 cm	7 cm

	V	B	H		V	B	H
(e)	45 cu m	5 m	$1\frac{1}{2}$ m	(f)	216 cu cm	6 cm	6 cm
(g)	140 cu cm	8 cm	$1\frac{3}{4}$ cm	(h)	99 cu m	3 m	$2\frac{3}{4}$ m

3 Find the width of a room which is 4 m long, 3 m high, and contains 42 cu m of air.

4 There are 3 cu m of water in a tank. If the tank is $1\frac{1}{2}$ m long and 1 m wide, what depth is the water? (Answer in cm.)

5 A beam contains $1\frac{1}{2}$ cu m of wood. If it is 200 cm in breadth and 100 cm thick, find its length.

6 A box has a volume of 192 cu cm. If it is 16 cm long and 4 cm high, what is its width?

7 The floor of a room which measures 5 m × 4 m is to be replaced. If 0·4 cu m of timber are needed, what is the thickness of the floor boards? (Answer in cm.)

8 A sheet of steel has a length of 20 cm, a breadth of 20 cm and a volume of 60 cu cm. What is its thickness?

VOLUME (2)

Now try these

Make a rough drawing whenever you can, and put in the measurements

1 Find the length of the third side in each of the following cubes or rectangular blocks, given the volume and the length of the other two sides:

	V	L	B
(a)	1331 cu cm	11 cm	11 cm
(b)	1425 cu cm	25 cm	19 cm
(c)	181·5 cu m	6 m	5·5 m
	V	L	H
(d)	3024 cu cm	21 cm	9 cm
(e)	200 cu m	10 m	4 m
(f)	$\frac{1}{100}$ cu m	4 m	2 m
	V	B	H
(g)	25 cu m	5 m	$1\frac{1}{4}$ m
(h)	5880 cu cm	$17\frac{1}{2}$ cm	14 cm
(i)	2115 cu m	45 m	10 m

2 A tank half full of water has a layer of ice on top of the water. If the tank is 2 m long and $\frac{1}{2}$ m wide, and the volume of the ice is 5000 cu cm, find the thickness of the ice.

3 A lorry contains 7 cu m of sand. If the load is 2 m wide and 1 m high, how long is it?

4 Twelve bricks of equal size together have a volume of 23 040 cu cm. If each brick is 24 cm long and 8 cm high, what is its width?

5 One cubic m of water measures 1000 litres. A tank holds 100 litres of water when full. If its internal measurements are 40 cm wide and 20 cm deep, what is its length?

6 A cupboard is 2 m high and 1 m wide. How deep (from front to back) must it be to have a capacity of one million cu cm?

GENERAL REVISION

First try these

Add

1 litres	ml	**2** $7\frac{1}{2} + 4\frac{2}{5} + 2$	**3**	£
23	421			40·17
6	35			120·90$\frac{1}{2}$
2	530			56·04$\frac{1}{2}$
+ 4	19			+ 70·38$\frac{1}{2}$

Subtract

4 m	cm	**5**	80 342	**6**	h min
67	45		– 37 685		20 39
– 23	94				– 3 47

Multiply

7	7·02	**8** kg	g	**9** $11\frac{1}{4} \times 6\frac{2}{5}$
	× 3·6	9	572	
			× 2	

Divide

litres   ml

10   57   4 ÷ 2   **11** 0·196 ÷ 0·14   **12** $16\frac{1}{2} ÷ 1\frac{5}{6}$

13 What decimal of 12·5 litres is 3·5 litres?

14 How many more cubic cm are there in a cube of side 0·5 m, than in a rectangular block 5 cm × 2 cm × 9 cm?

15 Write out the following example, and fill in the missing numbers: 2A9 + 31B + C45 = 1000.

16 A rectangular-shaped box has a volume of 98 cu cm. If the area of its base is 28 sq cm, what is its height?

17 Divide 50 pence between 2 children so that one has 14 pennies more than the other.

18 A traveller motors 220 km a day during his working week (5 days). If his car does 10 km to the litre, and petrol costs 15p per litre, what is his petrol bill for the week?

19 If $\frac{3}{5}$ of a sum of money is £1·20, what is $\frac{5}{8}$ of the money?

GENERAL REVISION

Now try these

Add

1		2	km	m	3 $\frac{1}{2} + \frac{3}{4} + 1\frac{5}{12}$

1
```
    198 307
     49 265
         94
  + 536 001
  ─────────
```

2
```
  km    m
   2   54
  16   87
  12   13
  +     8
  ─────────
```

3 $\frac{1}{2} + \frac{3}{4} + 1\frac{5}{12}$

Subtract

4
```
tonnes   kg
   32   540
 - 16   970
 ──────────
```

5
```
        £
  1021·15
 - 483·37½
 ─────────
```

6
```
   980·075
 - 273·948
 ─────────
```

Multiply

7
```
days    h
   4   19
      ×2
 ───────
```

8 $3\frac{4}{7} \times 2\frac{4}{5} \times \frac{3}{20}$

9
```
  293·765
     ×18
  ───────
```

Divide

10 £95·76 ÷ 9   **11** 9 kg 150 g ÷ 2   **12** 23·912 ÷ 5·6

13 A rectangle has an area of 70 sq m. If its length is 3500 cm, find its perimeter.

14 The rainfall for 8 months was 1·92, 2·03, 2·14, 1·68, 1·01, 0·95, 0·03, 0·84 in. What was the average monthly rainfall for this period?

15 Divide 750 into two parts so that one part contains $3\frac{1}{2}$ score more than the other part.

16 How many minutes are there from 12:07 on Wednesday until 23:58 on Thursday?

17 Divide £4·60 between 2 girls so that one gets £0·35 more than the other.

18 Seven-sixteenths of the number of children at a school were boys, while 288 were girls. How many boys were there in the school?

19 Five-sixths of a block of wood has a volume of 1440 cu cm. If the length of the block is 24 cm and its width is 18 cm, what is its thickness?

PROPORTION UNITARY METHOD

First try these

1 One litre of milk costs 10p. Find the cost of (a) $\frac{1}{2}$ litre, (b) 7 litres, (c) 8 litres.

2 Six dozen eggs cost £3·00. Find the cost of (a) 1 doz., (b) 4 doz., (c) 10 doz.

3 A man earned £8·00 for working 8 h. How much would he earn in (a) 1 h, (b) 5 h, (c) 12 h?

4 A map is drawn to the scale of 2 km to the cm. What length on the map stands for (a) 1 km, (b) 7 km, (c) 12 km?

5 A litre of water weighs 1 kg. Find the weight of (a) $\frac{1}{2}$ litre, (b) 5 litres, (c) 18 litres.

6 Two and a half tonnes cost £40·00. Find the cost of (a) 1 tonne, (b) 8 tonnes, (c) $\frac{1}{4}$ tonne.

7 A motorist travels 420 km in 10 h. How far did he travel on the average in (a) 1 h, (b) $2\frac{1}{2}$ h, (c) 20 min?

8 If 3 kg of tea cost £2·55, find the cost of (a) 1 kg, (b) $\frac{1}{2}$ kg, (c) $2\frac{1}{2}$ kg.

9 If 6 children use 12 litres, how much on the average could (a) 1 child, (b) 5 children, (c) 17 children use?

10 Four similar books weigh 7 kg. Find the weight of (a) 1 book, (b) 3 books, (c) 11 books.

11 A ship travels 300 sea miles in 12 h. How far does it travel in (a) 1 h, (b) $4\frac{1}{2}$ h, (c) 8 h?

12 Six kg of a certain substance costs £2·64. Find the cost of (a) 1 kg, (b) 11 kg, (c) 20 kg.

13 Eight men are together allowed £14·00 a day for expenses. How much is allowed to (a) 1 man, (b) 3 men, (c) 5 men?

14 Five rolls of wallpaper cost £4·50. What was the cost of (a) 1 roll, (b) 4 rolls, (c) 12 rolls?

15 Ten cm on a map stands for 10 km. What distance on the map stands for (a) 1 km, (b) 7 km, (c) 41 km?

16 If the fare for a bus journey of 20 km is 40p, find the fare for a journey of (a) 1 km, (b) 8 km, (c) 28 km, at the same rate.

PROPORTION UNITARY METHOD

Now try these

1 A dozen books of the same kind cost £8·88. Find the price of (*a*) 1 book, (*b*) 5 books, (*c*) 13 books.

2 A car travels 11 km in 10 min. How far will it travel in (*a*) 1 min, (*b*) 15 min, (*c*) 1 h?

3 Six cm on a map represent a distance of 300 km. Find what distance is represented by (*a*) 1 cm, (*b*) $2\frac{1}{4}$ cm, (*c*) $6\frac{3}{4}$ cm.

4 A cyclist rides 38 km in 2 h. If he goes at a steady speed, how far will he travel in (*a*) 1 h, (*b*) 30 min, (*c*) 5 h?

5 A gross of foreign stamps of the same value cost £3·60. What is the cost of (*a*) 1 stamp, (*b*) $9\frac{1}{2}$ dozen stamps, (*c*) 17 stamps?

6 A man earns £1300 in 26 weeks. How much will he earn in (*a*) 1 week, (*b*) 13 weeks, (*c*) 1 year?

7 4 litres of paint costing £10·80 is mixed with 4 litres costing £11·60. What is the cost of (*a*) 1 litre, (*b*) 4 litres, (*c*) 9 litres of the mixture?

8 A map is drawn to the scale of 5 km to the cm. What area is represented by (*a*) 1 sq cm, (*b*) $3\frac{1}{2}$ sq cm, (*c*) 10 sq cm?

9 A man saved the same amount each week. At the end of 5 years he had saved £1040·00. How much did he save in (*a*) 1 year, (*b*) 1 week, (*c*) 80 weeks?

10 A man travels to work by bus 5 days a week for 48 weeks in the year. During the year he travels 3840 km to and from work. How far does he travel (*a*) per week, (*b*) per day, (*c*) in 37 weeks?

11 An express train travels 595 km in $8\frac{3}{4}$ h, of which 15 minutes is spent in stopping at stations. Find the average distance the train travels in (*a*) 1 h, (*b*) 15 min, (*c*) $4\frac{3}{4}$ h, if the time spent at stations is not included.

12 An exhibition is open for 8 h a day and is visited during that time by 4320 people. Find the average number of people visiting it (*a*) each hour, (*b*) in 3 h, (*c*) in 1 min.

PROPORTION

First try these

1 If 10 boxes of sweets of the same kind cost £2·50, find the price of (*a*) 4 boxes, (*b*) 25 boxes.

2 If an aeroplane travels 4800 km in 6 h, how far will it travel in (*a*) 8 h, (*b*) $3\frac{1}{2}$ h, at the same speed?

3 The owner of a house with a frontage of 10 m pays £240·00 towards making the road. How much should a householder pay, at the same rate, for a frontage of (*a*) 8 m, (*b*) 13 m?

4 A car travelled 450 km on 30 litres of petrol. How far should it travel on (*a*) 20 litres, (*b*) 45 litres?

5 Twelve books cost £9·00. Find the cost of (*a*) 15 books, (*b*) 22 books, all at the same price.

6 Twenty children use 640 litres of water per day for drinking and washing. How much is required for (*a*) 9, (*b*) 37 children?

7 If 500 kg of coal is used in 4 weeks, (*a*) how long will 2 tonnes last, (*b*) how many tonnes will be used in 18 weeks?

8 A man earns £4·80 for working 4 h. How much will he earn in (*a*) 10 h, (*b*) 5 days, each of 8 h?

9 The return railway fare between two stations, 48 km apart, is £1·56. What would be the return fare between stations (*a*) 36 km apart, (*b*) 56 km apart?

10 Between midnight and 08:00 a steamer travels 144 sea miles. How far should it go between (*a*) 06:00 and 09:00, (*b*) noon and 17:00, if it travels at the same speed?

11 If a dozen cakes of soap weigh $1\frac{1}{2}$ kg, what would be the weight of (*a*) 27 cakes, (*b*) 45 cakes?

12 Two score of articles cost £5·00. Find the cost of (*a*) seven dozen, (*b*) a gross of these articles, if all cost the same.

13 A train went 320 km in 4 h. How long would it take to go (*a*) 96 km, (*b*) 200 km, at the same rate?

PROPORTION

Now try these

1 If 10 eggs cost 35p, find the cost of (*a*) 1 doz., (*b*) $3\frac{1}{2}$ doz.

2 If 3 sq m of linoleum costs £5·40, find the cost of (*a*) 7 sq m, (*b*) 25 sq m.

3 John can walk 8 km in $1\frac{1}{3}$ h. If he keeps up this speed, how long will it take him to walk (*a*) 12 km, (*b*) 15 km?

4 Twenty-eight kg cost £4·20. How much would (*a*) 72 kg, (*b*) 98 kg cost, at the same rate?

5 If 25 m of cloth can be bought for £20·00, how many m of the same material can be bought for (*a*) £56·00, (*b*) £42·50?

6 If 5 similar parcels weigh 8 kg, how many kg will (*a*) 11 and (*b*) 19 similar parcels weigh?

7 If 16 similar boxes hold 1440 objects of the same kind, how many such boxes are needed for (*a*) 450 and (*b*) 1080 of the objects?

8 An aeroplane travels 5100 km in 6 h. How far would it go, at the same average speed, in (*a*) 3 h 36 min, (*b*) 2 h 6 min?

9 If 37 toys can be bought for £44·40, find the cost of (*a*) 13, (*b*) 31 toys of the same kind.

10 A helicopter travels 288 km in 2 h 15 min. How far will it travel, at the same average speed, in (*a*) 40 min, (*b*) 1 h 15 min?

11 The average rainfall at a certain place is 19·6 cm in a month of 28 days. How much rain falls, on the average in (*a*) 17 and in (*b*) 23 days?

12 One cubic metre of substance weighs 60 kg. What is the weight of (*a*) 150 cu cm, (*b*) 10 000 cu cm? (Answer in g.)

13 On a map of scale 10 km to the cm a river is 2·75 cm long. How long will the river be on a map of scale (*a*) 4 km to the cm, (*b*) 50 km to the cm? (Think carefully.)

PROPORTION FRACTIONS

First try these

1 Three-quarters of a kg of chocolates cost £1·80. Find the cost of (a) $\frac{1}{4}$ kg, (b) 1 kg, (c) $2\frac{3}{4}$ kg.

2 Two-thirds of a lesson lasted 30 min. How long was the whole lesson? How long was $\frac{1}{4}$ of the lesson?

3 Find the value of 0·2 of a sum of money, if 0·5 of the money is equal to £10·00.

4 One and a half tonnes of sand cost £24·00. Find the cost of (a) $\frac{1}{4}$ tonne, (b) $\frac{1}{20}$ tonne, (c) $3\frac{1}{4}$ tonnes.

5 Five-eighths of a number is 10. Find (a) the number, (b) $\frac{3}{16}$ of the number, (c) $1\frac{1}{4}$ times the number.

6 A piece of timber 3·5 m long was $\frac{7}{10}$ of the longer piece from which it was cut. How long was the whole piece?

7 After cycling 8 km a man had completed $\frac{4}{15}$ of his journey. How far had he to go altogether?

8 Two and a quarter kg cost $22\frac{1}{2}$p. How much is that per kg?

9 Eighteen bottles of milk in a dairy turned sour. If this is $\frac{2}{5}$ of the total, how many bottles are there?

10 If £0·375 will purchase 3 articles, how many articles can be bought for (a) £1·00, (b) £1·125?

11 If 0·8 of a cake weighs 4 kg, find (a) the weight of the cake, (b) the weight of 0·25 of the cake, (c) the weight of 0·1 of the cake.

12 Three-tenths of a number is 8·4. Find the number.

13 If $1\frac{1}{12}$ of a sum of money is £4·18, find the sum.

14 Find the cost of $\frac{1}{2}$ m of material when $\frac{3}{8}$ m costs 45p.

15 An express train takes 7 min to travel $10\frac{1}{2}$ km. How long will it take, at the same rate, to travel 72 km?

16 If $\frac{7}{9}$ of a man's weekly wages comes to £28·00, how much does he earn per week?

17 When an aeroplane cruises at 660 km/h, it is going at $\frac{11}{12}$ of its normal speed. What is its normal speed?

18 Three and an eighth times a certain number is 75. Find (a) $\frac{1}{8}$, (b) $\frac{5}{16}$, (c) $\frac{1}{24}$ of the number.

PROPORTION FRACTIONS

Now try these

1 If $\frac{2}{5}$ of the cost of a new car is £600·00, what is its price?

2 Eleven-twentieths of a strip of metal in $4\frac{1}{8}$ m long. Find the length of (a) the whole strip, (b) $\frac{1}{3}$ of the whole strip.

3 When 0·004 of a number is equal to 17·2, find (a) the number, (b) 0·09 of the number.

4 If $\frac{5}{12}$ of the contents of a box weighs $7\frac{1}{2}$ kg, find the weight of $\frac{2}{9}$ of the contents.

5 Three-eighths of a house is valued at £2100·00. Find the cost of (a) the whole house, (b) a row of 8 similar houses.

6 A third of a sum of money exceeded 0·25 of it by £0·23. Find the value of the sum.

7 The cost of 100 g of a substance was £3·50. Find the cost of (a) $\frac{1}{8}$ kg, (b) 0·75 kg, (c) 7 g.

8 If $\frac{9}{16}$ of the distance between two towns is $38\frac{1}{4}$ km, how far apart are they?

9 A girl paid £17·22 towards her holiday, which was $\frac{7}{10}$ of the total cost. The remainder of the money was given her by two aunts. If one aunt gave her £4·00, how much did the other give?

10 Ninety-six is $\frac{3}{32}$ of a number. Find the value of twice the number.

11 After giving away $\frac{1}{6}$ of her money and spending $\frac{1}{3}$, Mary has 37p left. How much money had she at first?

12 Thirteen-sixteenths of a plank measures $3\frac{1}{4}$ m in length. Find the length of (a) $\frac{1}{8}$ of a plank, (b) 0·625 of a plank, (c) $3\frac{1}{4}$ planks.

13 If $\frac{7}{16}$ of a cask contains $1\frac{1}{2}$ litres more than $\frac{5}{12}$ of the cask, how much does (a) 1 cask, (b) $\frac{1}{9}$ of the cask hold?

14 A man sold $\frac{11}{15}$ of a plot of land for £550. How much should he get for the remainder if he sold it at the same price per square m?

15 Three-elevenths of a journey takes 1 h 44 min. How long should the whole journey take, travelling at the same average speed?

PERCENTAGES

First try these

1 Write as fractions the following percentages

(a) 5% (b) 90% (c) 40% (d) 75% (e) 20%
(f) $33\frac{1}{3}$% (g) 25% (h) 70% (i) $2\frac{1}{2}$% (j) 65%

2 Write as percentages the following fractions

(a) $\frac{1}{2}$ (b) $\frac{4}{5}$ (c) $\frac{1}{10}$ (d) $\frac{3}{4}$ (e) $\frac{1}{5}$
(f) $\frac{2}{3}$ (g) $\frac{1}{100}$ (h) $\frac{1}{20}$ (i) $\frac{3}{10}$ (j) $\frac{3}{5}$

3 Find the value of the following

(a) 25 per cent of 80, 32, £10·00, £4·00
(b) 80 per cent of 45, 50p, 1 tonne, 35 m
(c) 10 per cent of 73, £1·50, 60 min, 52 litres
(d) $33\frac{1}{3}$ per cent of 36, 90, 300 m, 9 km
(e) 75 per cent of 28, 1 kg, 40 sec, £1000·00
(f) 5 per cent of 60, 170, 70p, 8 km

4 What percentage of

(a) 48 is 12? (b) £26·00 is £13·00?
(c) 90 is 60? (d) 35p is 7p?
(e) 75 kg is 45 kg? (f) 1 tonne is 200 kg?
(g) $12\frac{1}{2}$ litres is $2\frac{1}{2}$ litres? (h) 200 is 30?
(i) 1 min 20 sec is 8 sec? (j) $\frac{1}{2}$ is $\frac{1}{4}$?

5 What percentage is

(a) 3p of 15p? (b) 250 g of 1 kg?
(c) 5p in the pound? (d) 80p in the pound?
(e) £10·00 of £5·00? (f) 100 g of 2 kg?
(g) $2\frac{1}{2}$ litres of 25 litres? (h) £400·00 of £100·00?

PERCENTAGES

Now try these

1 Write as fractions the following percentages

(a) 10% (b) $66\frac{2}{3}$% (c) 95% (d) $12\frac{1}{2}$% (e) 8%
(f) 30% (g) $1\frac{1}{4}$% (h) 47% (i) $\frac{1}{2}$% (j) 15%

2 Write as percentages the following fractions

(a) $\frac{1}{4}$ (b) $\frac{1}{25}$ (c) $\frac{1}{3}$ (d) $\frac{17}{100}$ (e) $\frac{9}{10}$
(f) $\frac{3}{8}$ (g) $\frac{1}{12}$ (h) $\frac{11}{20}$ (i) $\frac{3}{80}$ (j) $\frac{1}{50}$

3 Find the value of the following

(a) $2\frac{1}{2}$ per cent of 200, £100·00, 5 tonnes, 1 m 50 cm
(b) $66\frac{2}{3}$ per cent of 108, 450, 1 day, 30 km
(c) 30 per cent of 12·5, 204, 90p, 15 litres
(d) $12\frac{1}{2}$ per cent of 320, 0·64, £25·00, £750·00
(e) 20 per cent of 5400, 1·01, $\frac{1}{2}$ sq m, $2\frac{1}{2}$ km
(f) $8\frac{1}{3}$ per cent of 960, 66p, 40 min, 75 kg

4 What percentage is

(a) 0·5 of 2?   (b) 200 m of 2 km?
(c) 1000 of 1500?   (d) 144 cu cm of 1 cu cm?
(e) 30p of £10·00?   (f) 3·4 of 10·2?
(g) 36 min of 2 h?   (h) 10 g of 1 kg?
(i) £16·00 of £62·50?   (j) 20p of 50p?

5 What percentage of

(a) 45p is 135p?   (b) £3·00 is £4·50?
(c) $2\frac{2}{5}$ kg is 6 kg?   (d) 51 is 68?
(e) 2 is 2·25?   (f) 1000 ml is 17 ml?
(g) 390 is 429?   (h) £540·00 is £585·00?

PERCENTAGES PROBLEMS

First try these

1 Find the value of
   (a) 125 per cent of 64   (b) 175 per cent of 200
   (c) $133\frac{1}{3}$ per cent of 6 km (d) 120 per cent of 180
   (e) $102\frac{1}{2}$ per cent of 50 kg (f) 150 per cent of 62p
   (g) 110 per cent of 370   (h) $166\frac{2}{3}$ per cent of 15 kg

2 A girl obtained 29 marks out of 50 in arithmetic. What percentage was this?

3 A farmer had 350 birds of which 140 were ducks and the rest geese. What percentage of the whole were geese?

4 A man bought a television set costing £80·00 and paid 60 per cent down. How much had he still to pay?

5 In a class of 40 children 24 are girls. What percentage is this?

6 If 15 days in September had some rain, what percentage of the month was free of rain?

7 In a sale an article which usually costs £48·00 was reduced in price by 25 per cent. What was its selling price?

8 The captain of a football team scored 17 out of a total of 85 goals scored by the whole team during the season. What percentage of the goals did he score?

9 A boy had £1·50. His sister had 150 per cent of this amount. How much money had she?

10 In an examination 18 out of 54 pupils failed. What percentage passed?

11 A box holding 500 oranges was broken and 32 per cent were lost. How many oranges remained?

12 In a school of 250 children 4 per cent were away ill with measles. How many children were present?

13 Ten years ago a factory employed 300 people. Today it employs 225 per cent of this number. How many people does it now employ?

14 A man earns £2200 per year and saves $2\frac{1}{2}$ per cent of this. How much does he save in a year?

PERCENTAGES PROBLEMS

Now try these

1 (a) $112\frac{1}{2}$ per cent of 96 (b) 180 per cent of 1 m
  (c) $108\frac{1}{3}$ per cent of £7·20 (d) $101\frac{1}{4}$ per cent of 40p
  (e) 171 per cent of 1500 (f) $137\frac{1}{2}$ per cent of 7 litres
  (g) 450 per cent of 120 tonnes (h) $133\frac{1}{3}$ per cent of 2 days

2 A car is bought for £680·00 and 25 per cent of this is paid at once. How much remains to be paid?

3 There are 117 girls in a school and 143 boys. What percentage of the children on the school roll are girls?

4 If a man pays a bill of £30·80 within one month, $2\frac{1}{2}$ per cent is taken off. What would the reduced bill amount to?

5 The population of a village is $166\frac{2}{3}$ per cent of what it was 20 years ago. Then, 900 people lived in the village. How many people live there today?

6 Find the difference between 45 per cent of £17·00 and 20 per cent of £35·00.

7 What must be taken from 109 to make it equal to $8\frac{1}{2}$ per cent of 1200?

8 A salesman gets £10·50 for selling goods worth £126·00. Express his earnings as a percentage of his sales.

9 Five per cent of the total number of people living in a town is 1500. How many people live there altogether?

10 An express train used to take 7 h to make a journey. It now takes $3\frac{3}{4}$ per cent less time. How many minutes are saved on the journey?

11 In a sale, garments are reduced by $33\frac{1}{3}$ per cent. How much does Mother pay for garments originally priced at (a) £34·50, (b) 42p, (c) 96p?

12 Find the value of $\frac{1}{2}$ per cent of (a) 400, (b) £5000·00.

13 Fifteen per cent of 10 litres of milk turned sour. How much of the milk remained fresh?

14 Mary is 1·4 m tall and her brother Michael is 1·2 m. Express Mary's height as a percentage of her brother's.

PERCENTAGE INCREASE AND DECREASE

First try these

Note: Increase or decrease per cent is always calculated on the original amount: that is, the original amount is taken as 100%.

1 Find the percentage increase when
 - (a) 100 is increased to 136
 - (b) 50 is increased to 71
 - (c) 48 is increased to 60
 - (d) 80 is increased to 88
 - (e) £2·50 is increased to £5·00
 - (f) 45p is increased to 60p
 - (g) 150 is increased to 400
 - (h) 14·4 is increased to 16·2

2 Find the percentage decrease when
 - (a) 100 is decreased to 93
 - (b) 50 is decreased to 39
 - (c) 95 is decreased to 76
 - (d) 81 is decreased to 72·9
 - (e) £24·00 is decreased to £21·00
 - (f) £1·00 is decreased to 97½p
 - (g) 20·4 is decreased to 18·7
 - (h) 5p is decreased to 4p

3 Last year there were 500 children in a school. This year there are 580. What is the increase per cent?

4 A man's weekly wage was £33·60, but he now earns £42·00. What is the increase per cent?

5 An article which used to cost £2·40 now costs £1·60. What is the decrease in price, expressed as a percentage?

6 A rectangle had an area of 1 sq m. After a portion had been cut off, its area was 8500 sq cm. What is the percentage decrease in area?

7 The return railway fare to a certain town increased from £2·50 to £2·60. What was the increase per cent?

8 A house which cost £800·00 in 1939 now costs £8800. What is the increase per cent in the cost?

9 A shopkeeper bought an article for £1·50 and sold it for £1·75. What was his gain (that is increase) per cent?

10 The average number of dinners taken by the children of a school fell from 180 to 165. Find the decrease per cent.

PERCENTAGE INCREASE AND DECREASE

Now try these

Note: Increase or decrease per cent is always calculated on the original amount: that is, the original amount is taken as 100%.

1 Find the percentage increase when
   (a) 125 is increased to 150
   (b) 450 is increased to 600
   (c) 500 kg is increased to 0·830 tonne
   (d) £3·20 is increased to £4·00
   (e) 2·56 is increased to 2·88
   (f) 75 is increased to 93
   (g) 24p is increased to 30p
   (h) 1500 is increased to 1815

2 Find the percentage decrease when
   (a) 84 is decreased to 63
   (b) 900 is decreased to 750
   (c) 1020 is decreased to 918
   (d) 94·8 is decreased to 31·6
   (e) £1497·00 is decreased to £998·00
   (f) 25 kg is decreased to 22·5 kg
   (g) 1 tonne is decreased to 721 kg
   (h) 14 tonnes is decreased to 10·5 tonnes

3 One litre of water weighs 1 kg, while a litre of another liquid weighs 1057 g. How much heavier is this liquid than the water? (Express as a percentage.)

4 In 1937 a house was built for £875·00. The owner sold it in 1969 for £4375·00. What was the percentage increase in price?

5 A secondhand car dealer bought a car for £540·00 and had to sell it at £459·00. What was his loss per cent?

6 If a flask of water is heated from 20°C to 26°C, find the percentage increase in temperature.

7 The time required to make an article was reduced from 8 h 20 min to 7 h 45 min. What was the percentage saving in time?

8 What percentage of 0·63 is 0·84?

MISSING NUMBERS

All can try these

1 Add two terms to the following series

(a) 1, 4, 7, 10 — —

(b) 1, 5, 10, 16 — —

(c) 7, 10, 16, 25 — —

(d) 100, 90, 81, 73 — —

(e) 200, 195, 185, 170 — —

(f) (6 – 4), (9 – 6), (13 – 9), (18 – 13) — —

(g) 3, 6, 12, 24 — —

(h) 1, 7, 49, 343 — —

(i) $\frac{3}{2}$, $\frac{9}{4}$, $\frac{27}{8}$, $\frac{81}{16}$ — —

(j) $\frac{4}{3}$, $\frac{16}{9}$, $\frac{64}{27}$, $\frac{256}{81}$ — —

(k) 2048, 1024, 512, 256 — —

(l) $\frac{3125}{32}$, $\frac{625}{16}$, $\frac{125}{8}$, $\frac{25}{4}$ — —

(m) 81, 54, 36, 24 — —

2 Give the number indicated by the ? sign

(a) 2   5
6 13
4   9
8   ?

(b) 1   2
5 14
7 20
3   ?

(c) 20 15
8   6
12   9
10   ?

(d) 48 32
21 14
9   6
27   ?

UNEQUAL SHARING (2)

All can try these

1 Divide £5·00 between two children so that one gets four times as much as the other.

2 A cane is 108 cm long. Divide it into two portions so that one piece is twice as long as the other.

3 Divide 1 h into 2 periods so that one is 5 times longer than the other.

4 30p is to be shared between 2 girls so that one gets half as much again as the other. How much will each girl receive?

5 A sack containing 480 oranges is shared between two groups of children so that one group gets 3 times as many as the other. How many does each group get?

6 A man dies and leaves £1000·00. He leaves the money to his son, and a friend, so that the son gets 7 times as much as the friend. How much does the son get?

7 Twenty-seven tonnes of coal are delivered to two factories, one getting $1\frac{1}{4}$ times as much coal as the other. What weight of coal does each factory get?

8 Two churns of milk together hold 56 litres. One holds $2\frac{1}{2}$ times as much as the other. How much does the smaller churn hold?

9 Mr Jones bought two articles together costing £22·00. One article cost 1·75 times as much as the other. How much did each cost?

10 Divide 1 km into two sections so that one is 9 times longer than the other.

11 A house was sold for £7000 and the money shared between two sisters so that for every £3·00 one received, the other received £2·00. Find the share of each.

12 A duty of 11 h is divided between two guards so that one does $\frac{1}{5}$ longer than the other. How long is each guard on duty?

13 Two numbers added together equal 107·1. One of the numbers is 8 times the other. What are the numbers?

DISTANCE, SPEED AND TIME PROBLEMS

First try these

1 How long will it take a man to walk 14 km if he keeps up an average speed of $3\frac{1}{2}$ km/h?

2 A liner averaged 22·3 sea miles per hour. How far did it travel between noon and midnight?

3 A car travels 630 km in 9 h. Find its average speed.

4 An aeroplane travels at 560 km/h. How far will it travel in 2 h 12 min?

5 How far do I walk altogether if I walk for $1\frac{1}{2}$ h at 6 km/h, and for 2 h at $4\frac{1}{2}$ km/h?

6 A cyclist travels 51 km in 4 h 15 min. What was his average speed?

7 A car travels at the rate of 60 km/h. How far will it travel between 09:30 and 13:00?

8 Find the speed of a fighter aircraft which covers 232 km in 15 min.

9 Find the speed of a train, in m per sec, if it travels 15 km in 10 min.

10 Express the following speeds in m per sec
   (a) 108 km/h   (b) 36 km/h
   (c) 3 km per min   (d) 90 km/h
   (e) 720 km in a day   (f) $\frac{1}{4}$ km per min

11 Express the following speeds in km per hour
   (a) 15 m per sec   (b) 50 m per sec
   (c) 400 m in 1 min   (d) $\frac{1}{2}$ km in $2\frac{1}{2}$ min
   (e) $\frac{3}{4}$ km in 2 min   (f) 110 m per sec

12 A motor coach and bus start at the same time to go to a place 60 km away. The bus travels at 40 km/h and the coach at 60 km/h. How many minutes earlier does the coach arrive?

13 What is the speed of an aeroplane if it travels 2080 km between 09:50 and 13:05?

14 A police car chased thieves for 12 km, at an average speed of 54 km/h. How long did this chase take?

DISTANCE, SPEED AND TIME PROBLEMS

Now try these

1 In a walking race a man covered 8 km in 36 min. Find his average speed in km per hour.

2 A man's step is 76 cm. If he takes 110 paces per min, find the distance he would cover in 1 h. (Answer in km.)

3 Two groups of girls set out for a walk. Group A walks at 6·5 km/h for $3\frac{1}{2}$ h and Group B walks for 4 h at an average speed of 5·9 km/h. Which group walked the greater distance, and by how much?

4 A jet air liner covered a distance of 5250 km in $7\frac{1}{2}$ h. What was its average ground speed?

5 A girl leaves home at 08:30 and arrives at school at 08:47. If she walks at a speed of 66 m per min, find how far it is to the school. (Answer as a decimal of a km.)

6 If an aeroplane left London Airport at 10:56 and arrived at Belfast Airport (Northern Ireland) at 12:08, 468 km away, what was its average speed in km per hour?

7 A car covers 162 km in 2 h 15 min. Find (a) its speed and (b) how far it would go in 3 h 15 min at the same speed.

8 A Vickers Viscount aircraft maintained an average speed of 464 km/h over a distance of 1392 km. How long did it take to complete the journey?

9 A train covers 475·2 km at an average speed of 1·76 km per min. How long does it take on the journey?

10 A man walked 27·5 km at an average speed of 3·75 km/h. How long was he walking? (Answer in h and min.)

11 A bus leaves a town at 09:30 for a destination $187\frac{1}{2}$ km away and it travels at an average speed of 50 km/h. At what time should it reach its destination?

GENERAL REVISION

First try these

Add

1	8294	**2**	km	m	**3**	11·91
	35		3	135		0·073
	17 896		6	732		4·258
	+ 40 017		14	80		+ 29·46
			+ 29	265		

Subtract

4 $10\frac{1}{4} - 2\frac{5}{6}$

5	tonnes	kg	**6**	18 013
	21	450		− 956
	− 10	900		

Multiply

7	days	h	**8**	243	**9**	m	cm
	3	17		× 67		7	85
		× 2					× 2

Divide

10 $8149 \div 97$ **11** $19\frac{1}{5} \div 10\frac{2}{3}$ **12** km m 13 540 ÷ 2

13 Find the value of $\frac{3}{4}$ of $\frac{2}{3}$ of £5·22.

14 Twelve posts, each 1 m apart, are placed round the sides of a square plot of land. Find the area of the plot in square m.

15 What is the cost of 8·5 tonnes at £3·25 per tonne? Work throughout in decimals.

16 Divide £82·30 between two children so that one gets £4·20 more than the other.

17 A car was priced at £540·00, but this amount was increased by 5 per cent. What was the new selling price?

18 Jack owns $\frac{1}{2}$ a bungalow and his sister owns $\frac{1}{3}$ of it. If Jack's share is worth £1800, find the value of his sister's share.

19 If I can run at a steady speed of 100 m in 20 sec, how long will it take me to run round the edge of a rectangular field $84\frac{1}{2}$ m long and $40\frac{1}{2}$ m wide?

20 Reduce 51 litres to half-litres.

GENERAL REVISION

Now try these

Add

1 days	h	min
11	5	24
3	18	40
9	7	53
+ 2	16	1

2 $1\frac{9}{10} + 2\frac{1}{3} + 3\frac{1}{2}$

3 kg	g
12	190
3	251
17	129
+ 9	587

Subtract

4 m	cm
10	39
- 4	78

5	984 206
	- 519 485

6 $1\frac{1}{2} - \frac{7}{16} - \frac{1}{8}$

Multiply

7 £
13·91
× 8

8 kg	g
15	807
	× 2

9	48·61
	× 37

Divide

10 $312·4 \div 14·2$ **11** $28\frac{4}{5} \div 3\frac{3}{11}$ **12** 7 kg 800 g $\div 2$

13 Divide the year of 365 days into two portions so that the second is four times as long as the first. Then give the date of the last day of the shorter period, commencing on 1st January.

14 Find the average of 1·307, 49·2, 5·092, 61·86, 0·918, 25·307, 19·843.

15 A merchant bought goods to the value of £105·00 and sold them for £126·00. What was his gain per cent?

16 A large square is 9 times the area of a small square. What fraction of the perimeter of the large square is the perimeter of the small square?

17 Find the missing numbers AAA × AA = A22A.

18 If 10 sq m cost £22·50, how many square m can be bought for £15·75?

19 A train left at 09:45 and by 15:25 had covered 493 km. Find its average speed.

GRAPHS

All can try these

Always put a heading to your graph, and mark in the scales

1 Study the following data:

Day of week . . .	M	T	W	Th	F
Attendance in class . . .	42	39	43	42	38

Draw a graph to show how attendance varied with the day of the week.

2 Study the marks given below:

Subject .	Eng	Spell	Arith	Hist	Geog	Nature
Tom's mark	12	16	19	15	14	14

Draw a graph to show how marks varied with subject, and draw on your graph a horizontal line to show Tom's average mark.

3 A motorist left home at 08:00 and completed his journey at 19:00. He noted the number of litres of petrol in the tank and he drew this graph:

Graph to show how the amount of petrol varied with time.

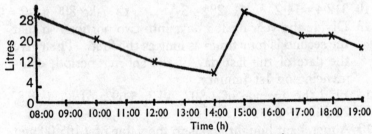

Study the graph and answer these questions:

(*a*) How many litres of petrol did the motorist use between 08:00 and noon?

(*b*) Can you suggest a likely reason why less petrol was used between noon and 14:00 than between 10:00 and noon?

(*c*) How many litres of petrol did the motorist use between 08:00 and 14:00, and between 15:00 and 19:00?

4 Draw a graph to show how the highest temperature (°C) reached during the day varied with the day of the week:

Day of week	. .	Sun	M	T	W	Th	F	Sat
Highest temperature	.	7	8	10	12	3	6	7

5 A cricketer scored the following number of runs in 8 innings:

Innings	. .	1	2	3	4	5	6	7	8
Runs scored	.	17	10	25	22	8	39	47	16

Draw a graph to show the variation in the number of runs scored with the innings.

6 The following graph shows the number of articles made in a factory during the twelve months of a year:

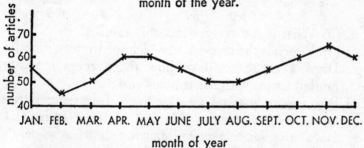

Graph to show the variation of the number of articles made with the month of the year.

Study the graph, and answer the following questions:

(*a*) How many articles were made each month?

(*b*) What was the average number of articles made monthly throughout the year, during the first six months, and during the last six months of the year?

(*c*) Can you suggest a reason why fewer articles were made in February?

(*d*) After production rose during April and May, it decreased again during June, July and August. Can you suggest why this was so?

(*e*) How many more articles were made during September, October and November, than during March, April and May?

MORE GRAPHS

All can try these

1 Some hens laid eggs as follows:

Day	Sun	M	T	W	T	F	Sat
Eggs	28	25	23	30	26	20	24

(*a*) On which day were the smallest number of eggs laid?

(*b*) On which day did the hens lay most eggs?

Draw a graph to show how the number of eggs laid varied with the day of the week.

2 The average monthly rainfall, in cm, for a certain town in the west of England is given by the table:

Jan	Feb	Mar	Apr	May	June
7·2	6·2	6·0	4·8	4·4	4·0

July	Aug	Sept	Oct	Nov	Dec
5·8	6·0	6·3	7·8	8·0	10·3

(*a*) What is the average monthly rainfall?

(*b*) Which is the driest, and which the wettest, month?

Draw a graph to show how the average monthly rainfall varies with the time of year.

3 Patients in hospital usually have their temperature taken in the morning (M) and in the evening (E). Here are some figures showing how a patient's temperature varied over 4 days. A normal temperature is 36·9°C.

Date	17th		18th		19th		20th	
	M	E	M	E	M	E	M	E
Temperature	37·5°	38·5°	37·9°	37·5°	37·1°	37·3°	37·0°	36·9°

(*a*) When was the patient's temperature at its highest?

(*b*) When did his temperature return to normal?

Draw a graph showing how the patient's temperature varied with time.

4 A baby was weighed regularly, and a record kept of his weight. The following information was obtained:

		At the end of					
Age	Birth	1 wk	2 wk	3 wk	4 wk	5 wk	6 wk
Weight	3·1 kg	2·9 kg	2·9 kg	3·0 kg	3·1 kg	3·3 kg	3·5 kg

Draw a graph showing how the weight of the child varied with his age. Then answer these questions:

(a) What happened to the weight of the child between birth and the end of the first week of life?

(b) What happened to his weight between the end of the 1st and the end of the 2nd week of life?

(c) When did the weight of the baby increase?

5 From Southwold to Northcote is 240 km. The line AB shows the journey made by motorist A, who left Southwold at 09:00 and reached Northcote at 17:00. The line CD shows the journey made by motorist C. Study the graph and answer the questions.

Graph to show the times of cars between Southwold and Northcote.

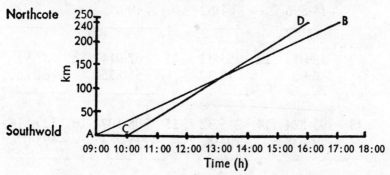

(a) What time does motorist C leave Southwold, and how long does he take on the journey?

(b) Find the average speeds of the two motorists.

(c) Which line has the steeper slope? What does this mean?

(d) How can you tell from the graph that motorist C overtakes motorist A?

(e) Find from the graph at what time C has covered 200 km. How far has A gone then?

6 Draw a graph to show how the population of England and Wales varied with time (1811–1951):

Date	1811	1831	1851	1871	1891	1911	1931	1951
Millions	10	14	18	23	29	36	40	43

REVISION OF DECIMALS

First try these

Set 1

	1		**2**		**3**		**4**
	0·34		3·96		16·84		41·72
	0·76		4·27		5·06		16·36
	1·29		0·81		7·13		0·09
	+0·58		+6·73		+12·59		+2·68

	5		**6**		**7**		**8**
	17·418		181·47		231·74		55·721
	93·215		29·53		16·82		104·215
	8·027		7·94		190·03		38·046
	+14·596		+1·62		+49·6		+1·927

	9		**10**		**11**		**12**
	12·07		54·32		67·914		137·95
	−6·98		−18·52		−56·875		−84·26

	13		**14**		**15**		**16**
	193·824		215·32		398·107		529·814
	−79·638		−194·87		−215·783		−495·076

Set 2

1 $0·073 \times 10$ **2** $1·296 \times 100$ **3** $9·84 \times 6$

4 $6·72 \times 3·4$ **5** $13·59 \times 0·7$ **6** $25·6 \times 4·3$

7 $3·74 \times 5·9$ **8** $0·19 \times 8·5$ **9** $0·067 \times 1000$

10 $7·7 \div 0·11$ **11** $19 \div 0·4$ **12** $1·12 \div 16$

13 $1·455 \div 0·485$ **14** $2·376 \div 3·3$ **15** $0·504 \div 12$

16 $26·676 \div 7·8$ **17** $4287 \div 1000$ **18** $102·01 \div 10·1$

REVISION OF DECIMALS

Now try these

Set 1

1	0·485	**2**	2·965	**3**	4·215	**4**	39·81
	0·129		4·017		9·768		4·9
	0·637		0·149		10·092		15·03
	+0·254		+1·688		+1·815		+27·62

5	93·12	**6**	271·38	**7**	312·85	**8**	729·831
	105·913		46·59		665·03		518·942
	16·407		105·75		297·91		463·015
	+200·01		+419·87		+5·64		+124·567

9	19·58	**10**	63·291	**11**	129·418	**12**	301·07
	− 13·74		− 14·886		− 36·057		− 190·84

13	837·215	**14**	1001	**15**	2175·004	**16**	10,000
	− 503·647		− 295·786		− 1368·890		− 4,975·38

Set 2

1 10·94 × 2·5   **2** 69·7 × 0·09   **3** 23·6 × 4·52

4 35 × 0·078   **5** 147 × 0·315   **6** 51·32 × 11·4

7 83·97 × 6·7   **8** 109·4 × 10·65   **9** 76 × 30·024

10 0·810 ÷ 0·009   **11** 1012 ÷ 100   **12** 0·728 ÷ 5·2

13 29·481 ÷ 31·7   **14** 2·226 ÷ 42   **15** 72·75 ÷ 1·94

16 862·5 ÷ 13·8   **17** 24·453 ÷ 0·09   **18** 96·336 ÷ 4·8

REVISION OF AVERAGES
AND VOLUMES

First try these

Set 1

Find the average of

1 32, 17, 16, 25, 29, 48, 60, 52. (Answer in decimal form.)

2 3·92, 4·07, 1·28, 5·36, 0·09, 10·01, 5·68, 9·04, 11·50, 6·19, 0·06.

3 £3·05, £4·55, £6·95, £5·15, £6·75, £9·85.

4 80 g, 50 g, 25 g, 60 g, 10 g, 35 g, 76 g.

5 5 km, 9 km, 13 km, 14 km, 6 km.

6 4·3 kg, 1·8 kg, 2·7 kg, 0·6 kg, 5·4 kg, 3·2 kg, 0·9 kg.

7 4 litres, 2 litres, 1 litre, 8 litres, 6 litres, 5 litres, 6 litres, 8 litres.

Set 2

1 Find the volume of the following cubes or rectangular blocks in which the length (L), breadth (B) and height (H) are given

	L	B	N		L	B	H
(a)	9 m	6 m	3 m	(b)	7½ m	3 m	1¼ m
(c)	2½ cm	2½ cm	2½ cm	(d)	3 m	1·5 m	0·75 m
(e)	4·7 cm	3·2 cm	1·6 cm	(f)	9¾ cm	7 cm	½ cm
(g)	11¼ cm	3⅕ cm	1½ cm	(h)	13 m	13 m	13 m

2 Find the length of the third side in each of the following cubes or rectangular blocks, giving the volume and the length of the other two sides:

	V	L	H		V	L	H
(a)	44 cu cm	11 cm	2 cm	(b)	180 cu m	15 m	4 m
(c)	96 cu cm	12 cm	4 cm	(d)	343 cu cm	14 cm	3½ cm

	V	B	H		V	B	H
(e)	242 cu cm	11 cm	2 cm	(f)	126 cu m	3½ m	2¼ m
(g)	1·98 cu m	1·1 m	0·9 m	(h)	17·55 cu cm	2·6 cm	1·5 cm

REVISION OF AVERAGES
AND VOLUMES

Now try these

Set 1

Find the average of

1 107·9, 384·2, 10·1, 236·5, 49·6, 97·8, 2·9.

2 £1·14, £4·47, £3·58, £5·00, £0·39, £0·26,
£2·92, £6·07, £3·80.

3 5·75 tonnes, 2·09 tonnes, 4·13 tonnes, 3·40 tonnes,
10·02 tonnes, 7·19 tonnes.

4 $\frac{1}{2}$, $\frac{1}{4}$, $\frac{3}{4}$, $\frac{1}{8}$, $\frac{3}{8}$, $\frac{5}{8}$, $\frac{7}{8}$, $\frac{1}{6}$, $\frac{5}{6}$.

5 3·27 m, 12·08 m, 8·14m, 9·86 m, 0·39 m, 13·53 m,
2·92 m.

6 9 days 20 h, 6 days 13 h, 10 days 7 h, 8 days 0 h,
4 days 22 h.

7 0·016, 0·257, 0·009, 0·184, 0·062, 0·976, 0·603,
0·015, 0·429, 0·010, 0·367, 0·048.

Set 2

1 How many cubic cm are there in (a) 9 cu m, (b) 10·25
cu m, (c) $15\frac{1}{2}$ cu m, (d) 23·6 cu m, (e) $27\frac{3}{4}$ cu m?

2 How many cubic cm are there in (a) 0·75 cu m,
(b) $\frac{7}{8}$ cu m, (c) $2\frac{1}{2}$ cu m, (d) 3·25 cu m, (e) 100 cu m?

3 Find the volume of the cubes and rectangular blocks
with the following measurements

	L	B	H		L	B	H
(a)	15 cm	14 cm	10 cm	(b)	8 m	7 m	3·125 m
(c)	$6\frac{1}{4}$ m	$6\frac{1}{4}$ m	$6\frac{1}{4}$ m	(d)	12·9 cm	10·6 cm	4·2 cm
(e)	$5\frac{1}{2}$ m	9 m	8 m	(f)	11·25 cm	12 cm	8 cm

4 Find the length of each of the following cubes or
rectangular blocks, given the volume and length of the
other two sides

	V	B	H		V	B	H
(a)	2744 cu cm	14 cm	14 cm	(b)	720 cu m	8 m	4·5 m
(c)	$1\frac{1}{2}$ m	2 m	5 m	(d)	10 648 cu cm	22 cm	22 cm
(e)	3456 cu cm	6 cm	$\frac{1}{8}$ cm	(f)	2538·9 cu cm	12·4 cm	10·5 cm

PROPORTION AND PERCENTAGES
PROBLEMS

First try these

1 Steamer tickets for 5 adults cost £12·50. Find the cost of (*a*) 1 adult ticket, (*b*) 7 adult tickets, (*c*) 1 child's ticket, reckoned at half the price of an adult's ticket.

2 On a hiking holiday 8 girls walk 84 km in 6 days. If they cover the same distance each day, how far do they walk in (*a*) 1 day, (*b*) 4 days, (*c*) 10 days?

3 Five dozen articles cost £3·60. Find the cost of (*a*) 3 articles, (*b*) a gross.

4 Ten sacks of flour weigh ¾ tonne. Find the weight of (*a*) 9 sacks, (*b*) 24 sacks.

5 Nine-sixteenths of a number is 2·7. Find the number.

6 Two-fifths of a m of cloth costs 48p. Find the cost of (*a*) ½ m, (*b*) 1¾ m.

7 Write as fractions the following percentages
    (*a*) 80% (*b*) 31% (*c*) 35% (*d*) 1% (*e*) 7½%

8 Find the value of the following
    (*a*) 60% of 30, 55, £4·50, 2 tonnes
    (*b*) 1¼% of 200, 10 litres, 20 km, 5 kg
    (*c*) 12½% of 1 min 20 sec, 168, 40 kg, 72p

9 What percentage is
    (*a*) 8p of 40p?   (*b*) 250 of 100?
    (*c*) 30 ml of 30 litres (*d*) 1·6 of 0·8?
    (*e*) 5 km of ½ km   (*f*) 500 m of 10 km?

10 Ninety pupils sat for a test and 81 gained more than half marks. What percentage of the pupils scored half marks, or less?

11 A shopkeeper prepared 40 litres of ice-cream for the Spring Holiday. He sold 34 litres. What percentage was this?

12 The fare for a steamer journey increased from £25·00 to £26·25. What was the increase per cent?

13 A man had a lawn 20 m × 20 m, but he dug up 50 sq m of it and planted flowers. What was the percentage decrease in the size of the lawn?

PROPORTION AND PERCENTAGES PROBLEMS

Now try these

1 Four dozen letters weigh 1 kg 200 g. Find the weight of (*a*) 1 letter, (*b*) 10 letters, (*c*) 38 letters, if each weighs the same amount.

2 In 12 weeks a boy's wages came to £198·00. How much does he earn in (*a*) 1, (*b*) 9, (*c*) 25 weeks?

3 If 9 equal bags of sand hold 630 kg between them, what weight of sand will (*a*) 31 bags, (*b*) 108 bags hold, if they are all of the same size?

4 Nineteen g of a substance cost $28\frac{1}{2}$p. Find the cost of (*a*) 39 g, (*b*) 57 g.

5 Four-fifteenths of a block of wood weighs 120·4 kg. Find the weight of (*a*) $\frac{1}{5}$ of the block, (*b*) $1\frac{1}{3}$ blocks.

6 Write as fractions the following percentages
   (*a*) 4% (*b*) $\frac{1}{4}$% (*c*) 56% (*d*) $16\frac{2}{3}$% (*e*) $37\frac{1}{2}$%

7 Find the value of the following
   (*a*) 40% of 2·5,   5 kg,   20 km,   8 min 20 sec
   (*b*) $\frac{2}{5}$% of 100, $12\frac{1}{2}$ litres,   25 kg,   900 sq cm
   (*c*) $3\frac{1}{2}$% of £1·00,   19·6,   2 kg.

8 What percentage is
   (*a*) 4p in the £?   (*b*) 160 sq cm of 1 sq m?
   (*c*) 37·5 of 62·5?   (*d*) 2 min 10 sec of 1 min?
   (*e*) 1 litre of 12 litres   (*f*) 9 m of 300 m?

9 A salesman was allowed $1\frac{3}{4}$ per cent of the value of the goods he sold. In one week he sold goods to the value of £2520·00. What were his earnings for that week?

10 A steamer normally did a journey in 6 days 16 h, but, on account of fog, it took $3\frac{1}{8}$ per cent longer. How long did the journey take?

11 What must be added to 311 to make it equal to $87\frac{1}{2}$ per cent of 500?

12 In a town of 3375 inhabitants $\frac{8}{9}$ per cent are ill with influenza. How many people is that?

13 What percentage of 35·76 is 31·29?

USING DIFFERENT SIGNS

All can try these

Set 1

1 $(6+5) \times 2$	**2** $6+(5 \times 2)$
3 $(12-4) \times 3$	**4** $12-(4 \times 3)$
5 $(10+8) \div 4$	**6** $10+(8 \div 4)$
7 $(20-15) \div 5$	**8** $20-(15 \div 5)$
9 $(\frac{1}{3}$ of $24)+6$	**10** $\frac{1}{3}$ of $(24+6)$
11 $(\frac{3}{4}$ of $16) \times 5$	**12** $\frac{3}{4}$ of (16×5)
13 $(8 \times 9)-(2 \times 4)$	**14** $8 \times (9-2+4)$
15 $17+11-25+10$	**16** $28-(7 \times 3)+6$
17 $(36 \div 9) \times 10$	**18** $36 \div (9 \times 10)$
19 $57-(24 \div 6)-2$	**20** $(57-24) \div (6-2)$
21 $(120-6) \times (4 \div 2)$	**22** $(42 \div 7)+(32 \times 4)$
23 $(144-4) \div 20$	**24** $(119+2-11) \div 10$

Set 2

1 $(\frac{1}{3}$ of $51p)+43p$	**2** $1\frac{1}{8}-(\frac{3}{4} \times \frac{8}{9})$
3 $(1 \cdot 96 \div 1 \cdot 4)+3-0 \cdot 6$	**4** £$0 \cdot 625+(\frac{1}{5}$ of £$1 \cdot 00)$
5 $(3\frac{1}{3} \times 1\frac{1}{5})-(\frac{1}{4} \times 3)$	**6** $(16 \cdot 4+7 \cdot 6) \div 12$
7 $1 \cdot 05$ kg $+(0 \cdot 15$ of 5 kg$)$	**8** $(1 \cdot 875-0 \cdot 5)$ of $20p$
9 $(\frac{5}{8}+\frac{3}{16}) \div 6\frac{1}{2}$	**10** $(2 \cdot 16 \times 4)-(10 \div 5)$
11 $7-(4\frac{1}{6}-2\frac{1}{9})$	**12** $(10 \times 9)-4-(3 \times 8)$
13 $(£10 \cdot 00-£6 \cdot 40) \div 4$	**14** £$20 \cdot 00+(£4 \cdot 38 \times 3)$
15 $4\frac{1}{2}-(\frac{1}{3}$ of $6)$	**16** $\frac{1}{2}-(\frac{1}{3}$ of $\frac{1}{4})+\frac{1}{5}$
17 $8(10-3)-5(6-4)$	**18** $26-\frac{1}{4}$ of (9×8)
19 $9 \times (32-27) \div 2\frac{1}{2}$	**20** $\frac{4}{11}$ of $(\frac{2}{3}+\frac{1}{4}) \div \frac{1}{6}$
21 $1 \cdot 75+\frac{1}{3}(1 \cdot 4+2 \cdot 6-0 \cdot 1)$	**22** $(8 \div 2)+(9 \div 3)-(10 \div 5)$
23 $67-(4-3)-(8 \times 5)$	**24** $(0 \cdot 25$ of $2 \cdot 8)-(0 \cdot 75$ of $0 \cdot 4)$

WORKING WITH LETTERS
INSTEAD OF NUMBERS

All can try these

1 Write down the number which is 1more than 3. Write down the number 1 more than x.

2 Write down the number 4 less than 6. Write down the number 4 less than y.

3 A boy had 12 marbles and won a more. How many had he then?

4 How many halfpence are there in (a) 1p, (b) 5p, (c) sp?

5 A girl is 10 years old. How old will she be in (a) 4 years' time, (b) in d years' time, (c) how old was she c years ago?

6 A boy had 11 pennies and he spent 5. How many had he left? Another boy had x pennies and spent y pennies. How many pennies had the second boy left?

7 How many cm are there in (a) p m, (b) q m?

8 Write down the sum of the three numbers a, b, c.

9 The width of a rectangle is w cm and its length is v cm. more. What is the length of half the perimeter?

10 If $x = 5$, $y = 3$, find the value of (a) $x+y$, (b) $x-y$, (c) $2x+2y$, (d) $3x-y$, (e) $1\cdot5x-2y$.

11 A girl is saving up £1·00. She has m pence so far. How much more must she save?

12 How many g are there in (a) 48 kg, (b) t kg, (c) k kg?

13 If $a = 2$, $b = 4$, $c = 6$, find the value of (i) $a+b+c$, (ii) $a+b-c$, (iii) $b+c-a$, (iv) $2a+b-\frac{1}{2}c$.

14 Multiply r by s and add t to your result.

15 A rectangle has a length of a cm and a width of b cm. What is its perimeter and its area?

16 A man earns £x per week. What will he earn in 12 weeks?

17 How far will a cyclist go in 4 h travelling at (a) p km/h, (b) u km/h, (c) z km/h?

18 Find how many lengths of string each x m in length can be cut from a ball of string y m long.

GENERAL REVISION

All can try these

Add

1   29 052
      6 479
    45 003
  + 98 140

2 $3\frac{3}{5} + 1\frac{5}{6} + 2\frac{2}{3}$

3

km	m
13	210
29	56
+ 8	947

Subtract

4   £
   624·73
  − 369·85$\frac{1}{2}$

5 10 000
  − 499·201

6

tonnes	kg
18	320
− 16	875

Multiply

7 $\frac{2}{7} \times 4\frac{1}{5} \times 6\frac{1}{4}$

8

litres	ml
17	768
	×2

9   97·6
  × 10·4

Divide

10 $5394 \div 93$ **11** £109·68 ÷ 12   **12** $3\frac{3}{8} \div 7\frac{5}{7}$

13 Write in figures seven hundred and fifty thousand and twelve.

14 A box contains £25·00 in coins of value 5p and 50p. If there are thirty-two 50p coins, find the number of 5p coins.

15 A hearth measures 1·5 m × 1·2 m. It is made of tiles 15 cm square. How many tiles are there?

16 Divide 36 kg of tea into 2 portions, so that one is 8 times as heavy as the other.

17 During a week in June a town had the following number of hours of sunshine: 11·3, 9·8, 8·7, 10·4, 9·3, 9·5, 11·7. What was the average number of hours of sunshine per day?

18 A cube has a side of 6 cm. What is its volume, and what is the total surface area of its faces?

19 Three-fifths of a stick is painted green, $\frac{1}{3}$ is painted blue, and the remaining 2 cm is unpainted. How long is the stick?